AF508673

Disentangling Mechanisms of Genomic Regulation of Cell Functions at the Gene Level

Disentangling Mechanisms of Genomic Regulation of Cell Functions at the Gene Level

Editors

Hans Binder

Arsen Arakelyan

MDPI • Basel • Beijing • Wuhan • Barcelona • Belgrade • Manchester • Tokyo • Cluj • Tianjin

Editors
Hans Binder
Interdisciplinary Centre for
Bioinformatics of Leipzig
University
Germany

Arsen Arakelyan
Institute for Molecular Biology,
Academy of Sciences of the
Republic of Armenia
Armenia

Editorial Office
MDPI
St. Alban-Anlage 66
4052 Basel, Switzerland

This is a reprint of articles from the Special Issue published online in the open access journal *Genes* (ISSN 2073-4425) (available at: https://www.mdpi.com/journal/genes/special_issues/Genomic_Regulation).

For citation purposes, cite each article independently as indicated on the article page online and as indicated below:

LastName, A.A.; LastName, B.B.; LastName, C.C. Article Title. *Journal Name* **Year**, *Volume Number*, Page Range.

ISBN 978-3-0365-0576-3 (Hbk)
ISBN 978-3-0365-0577-0 (PDF)

Contents

About the Editors

Hans Binder
ACADEMIC CAREER
1979 Diploma in Biophysics, Kharkov State University (Ukraina)
1984 Ph.D. (Dr. rer. nat.), Computer Experiments on Lipid Membranes, University of Leipzig (Germany)
2001 Habil. (Dr. rer. nat. habil.), Infrared Linear Dichroism of Ordered Amphiphilic Systems, University of Leipzig (Germany)
1985 – 2000: Lab head for optical spectroscopy and calorimetry, Leipzig University
2002–present Managing Director, Interdisciplinary Center for Bioinformatics University of Leipzig (Germany)

SCIENTIFIC INTERESTS
The development of bioinformatic methods for analyzing high dimensional omics-data in the context of cancer and other civilization diseases, applications in the broad context of systems biology, machine learning, and data sciences for biology and medicine

Arsen Arakelyan
ACADEMIC CAREER
1995–1999 Bachelor's degree in Biochemistry Specialization: Biochemistry Faculty of Biology, Yerevan State University Professional (Armenia)
1998–2011 Researcher, Laboratory of Macromolecular Complexes, IMB NAS RA
1999–2001 Master's degree in Biochemistry Specialization: Biochemistry Faculty of Biology, Yerevan State University (Armenia)
2001–2004 Ph.D. in Biology Specialization: Molecular and Cellular Biology IMB NAS RA
2008–2014 Deputy Director, IMB NAS RA
2011–present Group Leader, Research Group of Bioinformatics, IMB NAS RA
2014–present Adjunct Lecturer, American University of Armenia
2015–present Head of Bioinformatics Department, Russian-Armenia
2015–present Director, Senior Researcher IMB NAS RA

SCIENTIFIC INTERESTS
Bioinformatics; high-throughput data analysis; omics data analysis; pathway discovery and signal flow analysis; systems biology; functional genomics; telomere biology; data classification; marker discovery

 *genes* 

Editorial

Special Issue "Disentangling Mechanisms of Genomic Regulation of Cell Functions at the Gene Level"

Hans Binder [1,*] **and Arsen Arakelyan** [2]

[1] IZBI, Interdisciplinary Centre for Bioinformatics, Universität Leipzig, Härtelstr. 16–18, 04107 Leipzig, Germany
[2] Research Group of Bioinformatics, Institute of Molecular Biology NAS RA, 7 Hasratyan st, Yerevan 0014, Armenia; aarakelyan@sci.am
* Correspondence: binder@izbi.uni-leipzig.de; Tel.: +49-341-9716671; Fax: +49-341-16679

Received: 4 December 2020; Accepted: 5 December 2020; Published: 7 December 2020

The term "gene" was introduced more than a hundred years ago to define a "fundamental physical and functional unit of heredity" [1]. Its use refers back to Mendel's idea about discrete hereditable units, and it also serves as a reminiscence of Darwin's pangenesis theory. Watson and Crick's double helix discovery in 1962, which rationalized the molecular basis of replication and Crick's central dogma, claiming the one-directional information flow from DNA to proteins via RNA, conceptually represent milestones linking heredity with traits, and thus, genes with cellular function. From a practical perspective, after first determination of a gene sequence (coding bacteriophage MS2 coat protein, 1972), discoveries of Sanger sequencing (1977) and polymerase chain reaction (1983) mark technological milestones that enabled the decoding of the human genome until the end of the 20th century and, with the advent of expression microarray technology, the deciphering of the human transcriptome during the following decade (ENCODE project). Instead of solving the basic mysteries of life, these discoveries probably raised more questions about gene functioning than they answered. RNA appeared as a complex information flow machinery with highly specific nucleotide interactions fulfilling a large variety of functions in fine-tuning gene activity beyond transcription and translation. Moreover, DNA appeared not only as a one-dimensional textbook of the genetic code as thought previously. Instead, it was found to represent a highly complex three-dimensional polymer, where genetic information is managed via a multitude of conformations, interactions, and molecular components. Information management includes many tasks, such as reading, writing, erasing, repairing, transferring, and translating, where each of these steps must act in a proper way in space and time to ensure the functioning of life on cellular and organismal levels. Now, the central dogma appears less dogmatic because of the multidirectional nature of these interactions including feedback loops from the transcribed and translated parts towards their information source. Biological information management involves epigenetic mechanisms forming another feedback loop linking the phenotype and environmental factors back to genetics. Presently, diverse variants of next-generation sequencing technologies that have now reached single-cell resolution together with advanced metabolomics and proteomics methods provide an immensely powerful toolbox to discover mechanisms of gene functioning on the cellular level with increasing practical impact in healthcare and biotechnology. The latest cutting-edge examples from these areas are immuno-therapy against cancer (Nobel Price in Medicine 2019), CRISPR/Cas9 gene scissor (Nobel Price in Medicine 2020), and the mRNA vaccination technique just now becoming a sword against COVID-19.

This Special Issue collects seven publications addressing different topics around genomic regulation of cell functions at the gene level as examples illustrating various aspects of this field of discovery. Two original research publications deal with temporal aspects stored in the genome on completely different timescales; one on the scale of thousands of years and the other on the scale of minutes to hours [2,3]. Both works make use of similarity relations, with one considering the genomes of vine accessions [2] and the other considering the transcriptomes of single cells extracted from the flatworm [3]. The vine genome reveals "slow" mutational modifications, which enable reconstructing paths of distribution of wine agriculture and usage from the Middle East towards Western Europe on a long time-scale over many centuries. The single-cell transcriptome of the flatworm, on the other hand, reflects relatively "fast" changes of cellular programs upon differentiation of tissues proceeding on a much shorter time scale. Both applications illustrate the impact of another methodical ingredient, namely bioinformatics (also called computational biology), in order to process huge amounts of data generated by the novel high-throughput technologies and to "translate" them into useful (systems) biological information. Algorithmic developments, data science, and computational pipelines for effective practical use are inevitable parts of realizing "from-gene-to-cell-function" discoveries. Due to the size, and most importantly, the complex, often unknown intrinsic relations between the data, machine learning is an adequate approach to extract hidden information from the data. Both papers apply self-organizing map machine learning as a so-called "molecular portrayal" approach because data are reduced into handy dimensions and visualized in terms of easy-to-interpret images on an "individual" basis, e.g., for each measured unit (vine accession or worm single cell, respectively), making it an interesting approach for personalized medicine as well.

A topic with a medical impact has been presented in the paper of Filipenko et al. [4], who reported that the protective properties of a peptide drug (Semax) against ischemic stroke are associated with the compensation of mRNA expression patterns that are disrupted during ischaemic conditions. Leitao et al. [5] reviewed the geographic distribution of genetic variants of the CYP2D6 gene which are associated with different metabolization profiles, with a focus on Amerindian populations. This study underlines the impact of genomic variability on lifestyle factors and disease incidence in different ethnicities. In their in silico study, Chetta et al. [6] addressed mechanisms of the small noncoding RNA level of genomic regulation, a field that has risen in importance only in the last 15 years. Short sequence motifs of micro- and pi-RNA transcribed from non- (protein) coding regions of the DNA modify activity via binding to mRNA and transposons, and in this way, form intricate molecular networks between the different RNA-species and transcription factors with high impact for genomic regulation. Khokhlova et al. [7] reviewed DNA repair mechanisms in the early stages of mammalian embryonic development. In general, DNA repair links cell activity back to the genome because improper function causes errors in the genetic code to accumulate and result in the appearance of diseases and/or genetic drifts. Mutations that occur in somatic cells lead to dysfunction in certain tissues or organs, while a violation of genomic integrity during the embryonic period often leads to death. A mammalian embryo's ability to respond to damaged DNA and repair it, as well as its sensitivity to specific lesions, is not well understood. In their review, Lesne et al. [8] addressed a supramolecular aspect of genomic regulation, particularly the formation of nuclear bodies, membraneless organelles with crucial impact in regulating genome functions by promoting efficient interactions between distant genomic regions of the same or different chromosomes.

Overall, this collection of four original research papers and three reviews covers a series of mechanisms of genomic regulation and bioinformatics methods for their analysis; it provides examples and applications ranging from biotechnology to developmental biology to healthcare which will be of interest to researchers in different fields of molecular biology and medicine, agriculture, and computational biology.

Funding: This research received no external funding.

Conflicts of Interest: The authors declare no conflict of interest.

References

1. Noble, D. Genes and causation. *Philos. Trans. R. Soc. A Math. Phys. Eng. Sci.* **2008**, *366*, 3001–3015. [CrossRef] [PubMed]
2. Nikoghosyan, M.; Schmidt, M.; Margaryan, K.; Loeffler-Wirth, H.; Arakelyan, A.; Binder, H. SOMmelier—Intuitive Visualization of the Topology of Grapevine Genome Landscapes Using Artificial Neural Networks. *Genes* **2020**, *11*, 817. [CrossRef] [PubMed]
3. Schmidt, M.; Loeffler-Wirth, H.; Binder, H. Developmental scRNAseq Trajectories in Gene- and Cell-State Space—The Flatworm Example. *Genes* **2020**, *11*, 1214. [CrossRef] [PubMed]
4. Filippenkov, I.B.; Stavchansky, V.V.; Denisova, A.E.; Yuzhakov, V.V.; Sevan'kaeva, L.E.; Sudarkina, O.Y.; Dmitrieva, V.G.; Gubsky, L.V.; Myasoedov, N.F.; Limborska, S.A.; et al. Novel Insights into the Protective Properties of ACTH(4-7)PGP (Semax) Peptide at the Transcriptome Level Following Cerebral Ischaemia–Reperfusion in Rats. *Genes* **2020**, *11*, 681. [CrossRef] [PubMed]
5. Leitão, L.P.C.; Souza, T.P.; Rodrigues, J.C.G.; Fernandes, M.R.; Santos, S.; Santos, N.P.C. The Metabolization Profile of the CYP2D6 Gene in Amerindian Populations: A Review. *Genes* **2020**, *11*, 262. [CrossRef] [PubMed]
6. Chetta, M.; Di Pietro, L.; Bukvic, N.; Lattanzi, W. Rising Roles of Small Noncoding RNAs in Cotranscriptional Regulation: In Silico Study of miRNA and piRNA Regulatory Network in Humans. *Genes* **2020**, *11*, 482. [CrossRef] [PubMed]
7. Khokhlova, E.V.; Fesenko, Z.S.; Sopova, J.V.; Leonova, E.I. Features of DNA Repair in the Early Stages of Mammalian Embryonic Development. *Genes* **2020**, *11*, 1138. [CrossRef] [PubMed]
8. Lesne, A.; Baudement, M.-O.; Rebouissou, C.; Forné, T. Exploring Mammalian Genome within Phase-Separated Nuclear Bodies: Experimental Methods and Implications for Gene Expression. *Genes* **2019**, *10*, 1049. [CrossRef] [PubMed]

Publisher's Note: MDPI stays neutral with regard to jurisdictional claims in published maps and institutional affiliations.

 *genes*

Article

SOMmelier—Intuitive Visualization of the Topology of Grapevine Genome Landscapes Using Artificial Neural Networks

Maria Nikoghosyan [1,2,†], Maria Schmidt [3,†], Kristina Margaryan [4,5], Henry Loeffler-Wirth [3], Arsen Arakelyan [1,2] and Hans Binder [3,*]

1   Research Group of Bioinformatics, Institute of Molecular Biology of National Academy of Sciences RA, Yerevan 0014, Armenia; m_nikoghosyan@mb.sci.am (M.N.); aarakelyan@sci.am (A.A.)
2   Institute of Biomedicine and Pharmacy, Russian-Armenian University, Yerevan 0051, Armenia
3   Interdisciplinary Centre for Bioinformatics, University of Leipzig, 04107 Leipzig, Germany; schmidt@izbi.uni-leipzig.de (M.S.); wirth@izbi.uni-leipzig.de (H.L.-W.)
4   Research Group of Plant Genetics and Immunology, Institute of Molecular Biology of National Academy of Sciences RA, Yerevan 0014, Armenia; kristinamargaryan@ysu.am
5   Department of Genetics and Cytology, Yerevan State University, Yerevan 0025, Armenia
*   Correspondence: binder@izbi.uni-leipzig.de
†   Contributed equally to this study.

Received: 4 June 2020; Accepted: 15 July 2020; Published: 17 July 2020

Abstract: Background: Whole-genome studies of vine cultivars have brought novel knowledge about the diversity, geographical relatedness, historical origin and dissemination, phenotype associations and genetic markers. Method: We applied SOM (self-organizing maps) portrayal, a neural network-based machine learning method, to re-analyze the genome-wide Single Nucleotide Polymorphism (SNP) data of nearly eight hundred grapevine cultivars. The method generates genome-specific data landscapes. Their topology reflects the geographical distribution of cultivars, indicates paths of cultivar dissemination in history and genome-phenotype associations about grape utilization. Results: The landscape of vine genomes resembles the geographic map of the Mediterranean world, reflecting two major dissemination paths from South Caucasus along a northern route via Balkan towards Western Europe and along a southern route via Palestine and Maghreb towards Iberian Peninsula. The Mediterranean and Black Sea, as well as the Pyrenees, constitute barriers for genetic exchange. On the coarsest level of stratification, cultivars divide into three major groups: Western Europe and Italian grapes, Iberian grapes and vine cultivars from Near East and Maghreb regions. Genetic landmarks were associated with agronomic traits, referring to their utilization as table and wine grapes. Pseudotime analysis describes the dissemination of grapevines in an East to West direction in different waves of cultivation. Conclusion: In analogy to the tasks of the wine waiter in gastronomy, the sommelier, our 'SOMmelier'-approach supports understanding the diversity of grapevine genomes in the context of their geographic and historical background, using SOM portrayal. It offers an option to supplement vine cultivar passports by genome fingerprint portraits.

Keywords: grapevine genomes; genetic diversity; dissemination of vine; genome passporting; self-organizing maps; genome portrayal

---

1. Introduction

Grapes are not only tasty, but also one of the most economically and culturally important crops. They are used for both winemaking and fresh ('table') consumption. Grape is unique, not only because it is a major global perennial crop, but also because of its historical and cultural connections with the development of humans. According to the OIV (International Organization of Vine and Wine) the surface area of the world vineyard is estimated at 7.4 million hectares, with a production of 76 million tons of fresh grapes and 260 million hectoliters of wine (http://www.oiv.int/). Grapevine (*Vitis Vinifera*) is one of the oldest of the cultivated plants for which living progenitors still exist. The broad geographic area of the wild grapes and a bewildering number of different forms expanding from Europe to Asia and Caucasus continue to intrigued researchers, with a crucial question about the origin and domestication of the grape. Archaeological and historical studies suggested that the cultivation of the domesticated grape (*V. vinifera* L. subsp. sativa) started about 10,000 to 8000 years ago, from its putative wild ancestor (*V. vinifera* L. subsp. sylvestris), and the primary center of domestication was located between the Near East [1] and the Transcaucasian region [2]. These regions were populated by the Neolithic Shulaveri-Shomu, and later Kura-Araxes cultures, in the South Caucasus and Fertile Crescent regions, ranging from (today's) Georgia, through Azerbaijan, Armenia, northern Iran and eastern Anatolia [3]. Later, grapevine disseminated into the Southern Balkans and East Mediterranean Basin, then to the Western Europe and, finally, domesticated grapes were introduced to Central Europe during the first millennium BCE [4]. Hereby, wine grapes have followed human civilization from South Caucasus, southwards and westwards, spread first by seafaring Phoenicians and Greeks, and later affected by Roman and Ottoman Empires throughout the Mediterranean world [5]. During its spreading across the regions, grapes slowly mutated and adapted to their environments. The slow divergence over thousands of years in combination with the spontaneous hybridization, somatic variation and selection by humans created the incredible diversity of the more than 6000 cultivated varieties, which, in contrast to its wild progenitor, is more diverse and heterozygous according to OIV (http://www.oiv.int/). Different authors evidenced the presence of secondary domestication centers, where spontaneous hybridizations among wild plants and cultivated forms or targeted selection, created the pattern of the modern Western European cultivars [4]. M.A. Negrul, an outstanding pioneer researcher of vine, subdivided the varieties based on geographic origin and morpho-ecological traits into the ecotypes occidentalis (France, Germany, Spain, Portugal), pontica (Asia Minor, Romania, Hungary, Greece, Georgia, Bessarabia) and orientalis (Armenia, Iran, Afghanistan, Azerbaijan and Central Asia), where the primary center of a given plant is the region of the richest genetic diversity [6].

In the last years, whole-genome studies of vine cultivars using genotyping and sequencing technologies have added novel knowledge about the diversity, geographical relatedness, historical origin, phenotype associations, genetic markers and distribution paths of the vine [4,7–11]. Genetic relatedness between cultivated (*V. Vinifera* L. subsp. sativa) and wild (*V. Vinifera* L. subsp. sylvestris) grapes suggest at minimum two separate domestication events, one derived from the Transcaucasian wild grape and another one in Western Europe, where cultivars experienced introgression from local wild grape [4,7]. Associations between candidate genes and important agronomic traits, such as berry shape and aromatic compounds, provide possible genetic targets for grapevine improvement [9]. Breeding for larger berries has been related to genetic signatures of positive selection for table grapes in geographic regions, where alcohol was prohibited by religious rules [8]. The further impact of genomic studies will depend on the understanding of the genotype-phenotype associations, especially for complex traits governed by polygenic architecture, genotype-environment interactions in different geographic regions [11]. These tasks challenge activities and research infrastructure for collecting cultivar accessions, sequencing and data deposition, for providing phenotype-genotype databases and vine passporting projects, and, last but not least, the development of bioinformatics methods and tools which enable an analysis of these data, in terms of knowledge mining and feature extraction, in an easy and intuitive fashion.

Here, we apply SOM (self-organizing maps) portrayal [12], a neural network-based machine learning method with strong visualization capabilities to genome-wide Single nucleotide polymorphism (SNP) data of nearly eight hundred grapevine cultivars collected from Middle Asia in the East to Iberian peninsula in the West and from overseas regions [11]. SOM portrayal has been developed by us for the detailed analysis of high-dimensional omics data, including diversity and developmental issues, feature selection, knowledge mining and phenotype association of transcriptomic [13–15], epigenetic [16,17], proteomics [18] and genetic data [19], and of combinations of them [20]. In the context of plants, SOM-portrayal has been previously applied by us for the typing of algae of the genus Prototheca [21] and by others for studying early seed maturation in garden pea [22]. This work aims at providing a prototypic application of SOM machine learning and linked downstream analytics to illustrate its potency for studying plant genomes.

Strengths of the applied method are dimension reduction and visualization capabilities [23]. Particularly, our method generates accession-specific images of the data landscapes. These personalized omics-portraits provide options for the intuitive evaluation of feature space, and for mutual comparisons between the individual accessions. The recent application of SOM-portrayal to population-level distributions of disease-related human SNP-variants demonstrated its capability to extract genetic features and to describe genetic diversity, based on the topology of the SNP-landscapes [19].

This study aims at applying SOM-portrayal to the genetics of vine, in order to generate SNP-landscape of grape genotypes, to relate its topology to the geographical distribution, to possible paths of cultivar distribution, to selected genetic markers, and link them to grape utilization. In gastronomy, the sommelier is the wine 'waiter' (or 'waitress'), who recommends combining wine with food based on traditional knowledge about the character of wine varieties, their taste and geographic origin, and thus to support the decision-making of guests for a successful dinner. In analogy, our 'SOMmelier'-approach aims at supporting understanding genetic diversity and relatedness of grapevine genomes in the context of their geographic and historical background by application of SOM portrayal.

2. Materials and Methods

2.1. Data and Preprocessing

Grape cultivars data: grapevine genetic SNP- and phenotype data were taken from [11]. The data set consists of 783 grapevine samples originated from 4 grapevine collections (Table S1, Passport data of 783 cultivars included in the study) collected from eleven geographic regions, ranging from Middle Asia to Iberia and New World accessions. The genotype data matrix of 10,207 SNPs for 783 unique samples was taken from https://urgi.versailles.inra.fr/Species/Vitis/Data-Sequences/Genotyping-data repository.

Human population data: For a comparison of genomic relatedness between worldwide geographic strata, we analyzed genome-wide microarray SNP data (Illumina 650Y arrays) from the Human Genome Diversity Project (HGDP) [24], as described previously [19]. This data consists of 650,0000 SNPs for 940 individuals from 8 geographical regions (Africa, Europe, Middle East, South and Central Asia, East Asia, Oceania, and America). Genotypes for 99 individuals of Armenian ethnicity (Illumina Human Omniexpress microarray platform) were taken from a publication by Haber et al. [25]

2.2. Conceptual Overview

Our analysis concept is based on the following main ingredients: (i) transformation of SNP data into excess minor allele frequencies (eMAF), which increases sensitivity for subtle differences between the genomes of the vine accessions; (ii) dimension reduction of the more than 10,000 SNP data into a matrix of 2500 so-called 'meta-SNPs', using unsupervised clustering and their visualization in terms of individual accession portraits using SOM machine learning. (iii) further reduction of dimension by segmentation of the SNP-portraits into about one-to-two dozen so-called spot clusters of correlated SNPs, which serve as fingerprint features of the vine genomes; (iv) diversity analysis of accessions using different approaches (principal component analysis, correlation networks, t-SNE, minimum spanning tree). Each of the approaches used enables discovering different aspects of the mutual relatedness between the cultivars. (v) So-called pseudotime analysis as a special type of diversity analysis to analyze genetic similarities in terms of multibranched trees with possible impact for temporal diversification of vine cultivars; (vi) phenotype associations to map information of vine utilization on the SNP landscapes.

2.3. Allele Coding and SNP-Score

In order to further process the data, we coded the genotype of each SNP using a trinary code, with '0' for homozygous major alleles (AA), '1' for heterozygous alleles (Aa and aA), and '2' for homozygous minor alleles (aa).

First, let us consider one SNP in the data set of N cases, e.g., of N =783 cultivar samples. The fractions of the three genotypes of this SNP are defined as $p_{ij} = N_{ij}/N$ (i,j = A, a), where N_{ij} is the number of SNPs with the respective allele in the data set. The minor allele frequency of a SNP is $MAF = 2\, p_{aa} + 1\, (p_{Aa} + p_{aA})$ with $p_{aa} < p_{AA}$. With the SNP-code introduced above one finds for its mean value averaged over all cases, <SNP-code> = $p_{AA} * 0 + (p_{Aa} + p_{aA}) * 1 + p_{aa} * 2 = 2\,MAF$, i.e., the mean SNP-code of a SNP equals twice its MAF. For further data processing, we define a SNP-score by centralizing the SNP-code of each SNP with respect to its mean value, SNP-score = SNP-code − 2 MAF. One obtains SNP-score = −2 MAF for the major allelic SNPs, SNP-score = 1 − 2 MAF for heterozygous SNPs and SNP-score = 2 (1 − MAF) for the minor allelic SNPs.

Second, let us consider a group of n correlated SNPs in the data set, and calculate the mean of their SNP-score in one selected sample. In the next subsection, we show that such a group of SNPs is given, e.g., by a meta-SNP, or a 'spot-module' as obtained below in SOM-analysis. Analogous consideration as above deliver the result for the group-averaged SNP-score, <SNP-score> = 2 (MAF - <MAF>) where < … > here denotes group averaging referring to one selected sample. Hence, the group-averaged SNP-score estimates the deviation of the mean MAF of the considered SNPs, in a certain sample from their mean in the considered population. Accordingly, it can be understood as an excess MAF-value (eMAF), where positive SNP-scores mean higher frequencies than in the population while negative values refer to reduced frequencies. The centralization of each feature (here the SNP-code of each SNP) with respect to its mean value averaged over all samples is applied as a standard preprocessing step in SOM analysis. Features space of centralized values is more sensitive to subtle differences between samples than the feature space of non-centralized values.

2.4. SOM Portrayal and Spot Detection

SOM-portrayal of SNP data was performed as described previously [19]. In short: our SOM implementation used a ternary code with the values 0, 1 and 2 for major homozygous, heterozygous and minor heterozygous genotypes, respectively, as introduced above. Next, SNP data were mean centralized and then clustered using the self-organizing map (SOM) machine learning. SOM training translates the original data matrix consisting of the allele scores of N = 10,206 SNPs collected from M = 783 cultivar accessions into a data matrix of reduced dimensionality of K = 2500, so-called meta-SNP profiles. Hereby, the term 'profile' denotes the vector of eMAF score values across the cultivars. The SOM training algorithm distributes the SNPs over the K micro-clusters of meta-SNPs, by minimizing the *Euclidean* distance between the SNP- profiles as a similarity measure. It ensures that SNPs with similar profiles cluster together in the same or in closely located meta-SNPs. Each meta-SNP profile can be interpreted as the mean profile averaged over all SNP profiles of the respective meta-SNP cluster. For each cultivar accession, the meta-SNP score obtained provides the excess MAF (eMAF, see above). It is defined as the difference between the mean minor allele frequency (MAF of meta-SNP) of all SNPs collected in this meta-SNP in this particular sample, minus the mean MAF of these SNPs averaged over all samples. The eMAF values of each cultivar accession are visualized by arranging them into a two-dimensional M = 50 × 50 grid, and by using a red to blue color-code for maximum to minimum eMAF-values in each of the grid images. These images 'portray' the genetic landscape of each accession studied in units of the eMAF SNP-score. We used SOM as implemented in the "oposSOM" R package [23]. Cultivars were labelled according to the geographical region where they were selected (see [4] for details) and, alternatively, using different cluster assignments (see below).

Mean SNP-SOM portraits of cultivars from the same geographic regions were obtained by averaging the meta-SNP values of the respective individual SNP-portraits. The self-organizing properties of the SOM algorithm generates red spot-like regions referring to correlated SNP-profiles showing high eMAF-values in the respective accession portraits. We used segmentation algorithms developed previously [23] to extract so-called spot-clusters from these (red) regions. Each of these spot-clusters includes hundreds of SNPs.

2.5. Phenotype Association, Accession Diversity, Pseudotime Analysis and Clustering

Categorical phenotypes such as the utilization of grapes for table or wine usage were associated with the SOM SNP-landscape by performing ANOVA (analysis of variance) of SNP-metagenes of the sub-collection of cultivars of a certain phenotype (e.g., of all cultivars of table-vine usage) and coloring the meta-SNP pixels in the SOM according to the obtained p-value (in units of -log(p)), from red (high) to blue (low). Phenotype correlation maps were obtained by calculating the point biserial correlation between the eMAF profile of each metagene and the respective phenotype profile, which is given, e.g., for 'table vine utilization' by categorical values '0' (no table usage) and '1' (table usage). The metagenes of the map were then colored between red (maximum correlation) and blue (minimum correlation). Point serial correlation de facto provides difference portraits between cultivars of the respective phenotype and all others, e.g., between the mean portrait of cultivars of wine utilization, minus the mean portraits of cultivars of non-wine utilization.

Accession diversity analysis was performed based on meta-SNPs using principal component analysis (PCA), similarity net and minimum spanning tree plots, based on correlation metrics between the individual SOM-portraits as implemented in oposSOM [23]. t-SNE (t-distributed Stochastic Neighbour Embedding), URD-plots and pseudotime plots were generated using the program 'URD' [26]. URD estimates multibranched developmental trajectories based on mutual similarities between the accessions, a k-nearest neighbor graph presentation and directed (from 'root' to 'tip') simulation of a diffusion-like process. It provides a pseudotime (PT) value between zero and unity for each accession, where values near zero mean closer similarity to the root accessions, and values near unity mean closer similarity to tip-accessions. URD generates branched tree structure by joining diffusion trajectories passing through the same accessions.

Grapevine cultivars were stratified according to the geographic region in agreement with [10], where they were collected and using an eight cluster C1–C8 (C-clusters) division, according to [11]. Independent 't-SNE' clustering was performed using the Jaccard-algorithm implemented in 'URD' software [26], which provided ten clusters (see below).

3. Results

3.1. SOM Portrayal of Cultivars among Geographical Regions

SOM-training of the SNP-data of 783 grapevine cultivars provided a SNP-portrait of each accession. Each portrait represents a two-dimensional pixel-plot, where each of the pixels, the so-called meta-SNP, is colored according to its excess minor allele frequency (eMAF) of correlated meta-SNPs in the chosen cultivar (Additional File 1). According to their origin, the SNP-portraits were stratified into nine geographic regions, ranging from Middle and Far East (MFEA), Eastern Mediterranean and Caucasus (EMCA) over Russia and Ukraine (RUUK), Balkan (BALK), Western and Central Europe (WCEU), Italian Peninsula (ITAP) to Maghreb (MAGH) and Iberia (IBER) in the 'Old World' (Figure 1A). Cultivars from Armenia (ARM) and from the New World (NEWO), including America, Australia, New Zealand and South Africa, were separately considered (see Abbreviations for glossary of geographic regions). The mean SNP-portrait of each region reveals specific patterns of red spot-like areas which visualizes the respective genotype in terms of the 'meta-SNP' score landscape, where red/blue areas refer to positive/negative eMAF values, meaning that the mean minor allele frequency (MAF) of the SNPs in this spot from this region exceeds/falls below their mean MAF value averaged over all cultivars studied of all geographic regions (Figure 1B). The diversity of cultivars from Armenia and the portrait averaging applied is illustrated in terms of a hierarchical clustering tree (Figure 1C). The vine accessions roughly split into three major groups in the similarity net (Figure 1B), namely into cultivars from WCEU (and ITAP), from IBER and from the other regions, where the former two groups distribute along the first two principal component PC1 and PC2 of principal component analysis, while PC3 segregates cultivars of BALK from RUUK, MFEA and EUCA (Figure 1D).

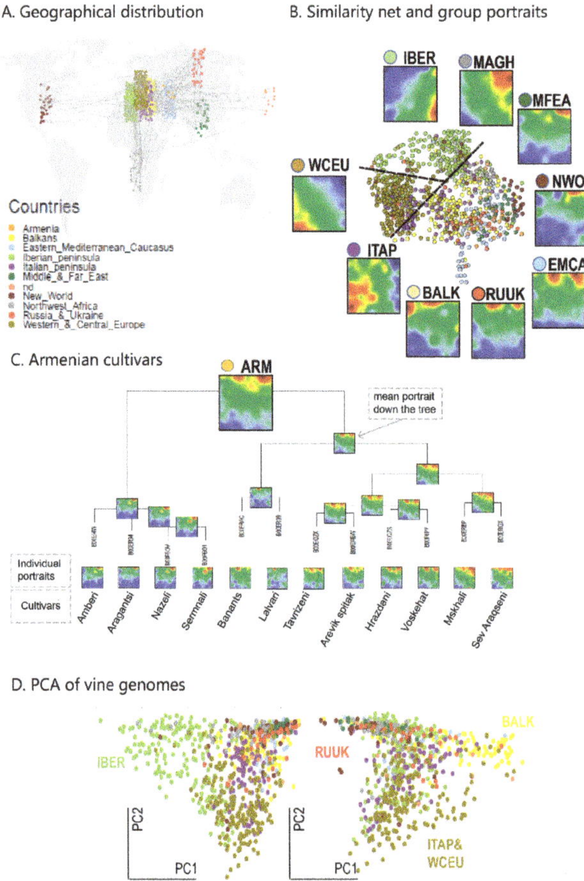

Figure 1. Self-organizing maps (SOM) portrayal of grapevine Single Nucleotide Polymorphism (SNP)-data: (**A**) Sampling of grapevine cultivars from different geographic regions. The thin lines visualize minimum spanning relations between the samples. (**B**) SOM-portraits of cultivars from different geographic regions indicate specific spot patterns of SNPs. The similarity net shows three major clusters of cultivars originating from the Iberian Peninsula (IBER), western and central Europe (WCEU) and Italian Peninsula (ITAP), and the other accessions originating mostly from Eastern Europe, Maghreb, Near East and Middle Asia (see dashed lines). (**C**) Hierarchical clustering tree of the cultivars originated from Armenia. The SOM portraits were progressively averaged in upwards direction. The mean portrait shows the weighted average of the spot patterns of the individual portraits. (**D**) Principal component plots reveal the distribution of accessions from WCEU (and from the Italian peninsula, ITAP) and IBER, along the first two principal components PC1 and PC2, respectively, and the split of the samples from Russia and Ukraine (RUUK) and the Balkan (BALK) along PC3. Color code of samples is provided in part B.

3.2. The Topology of SNP-Landscapes Characterizes the Diversity of Cultivars

The inspection of the individual accession portraits (Additional File 1) reveals a high diversity of textures. They reflect the allelic landscapes in terms of areas of positive and negative eMAF values, colored in red and blue, respectively, referring to enriched or depleted minor allelic genotypes. The spot summary map collects the 'red' spots most frequently observed in the individual maps (Figure 2A). It provides the SNP-landscape of the accessions studied by distributing the SNPs across the x-y-plane, making use of the self-organizing properties of SOM machine learning, and then, plotting the respective eMAF values along the 'height' dimension. The topology of this landscape is given by the eMAF-height profile. Overall, we identified fifteen such spots labelled with capital letters A–O. Overall, they agree with the spot patterns observed in the mean portraits of the geographic regions considered (Figure 2B). Lists of SNPs collected in the spots are provided in Additional File 2. The 'variance map' indicates that the spot regions contain SNPs of the largest variance of their eMAF SNP-score among the cultivars. The maps can be divided into three major regions of spots observed predominantly in the SNP-portraits of IBER, WCEU and 'the East' (from Central Europe), respectively (Figure 2A). The spot frequency distributions of the different geographical regions clearly reflect these three different patterns, in terms of bar plots centered around spot G frequently expressed in grapevine cultivars from IBER (see the vertical dashed line for orientation). In contrast, cultivars from WCEU and ITAP frequently express spots H–J (to the right from G), while grapevine cultivars from the 'East' frequently express spots A–F located to the left from spot G. BALK and RUUK cultivars reveal intermediate patterns. Most SNPs, namely between 300 and 550 per spot, were included in spots E–H, while each of the remaining ones contains about 200 SNPs.

The spot-profiles show the mean values of the eMAF-scores averaged over all meta-SNPs included in the respective spot among the cultivars and, particularly, their region-specific alterations (Figure 2C). Largest SNP-score values, and thus high excess frequencies of the minor alleles included in the spot, were observed for WCEU and ITAP, followed by IBER and the 'East' (see boxplots below the heatmap in Figure 2C). Hence, one finds a tri-partition of accessions with respect to the amplitude of the SNP-score as found also for spot-distributions. The variance of the SNP-score (and entropy, which estimates the degree of randomness of spot patterns) is largest for WCEU and IBER, and to a less degree MFEA and EMCA, paralleled by the largest number of spots per accession, reflecting the largest genetic diversity in these regions (see also the spot-number distribution in the last row of plots in Figure 2C), which are assumed to serve as criteria of centers of vine cultivation [6]. The correlation map shows that SNP-scores are strongly anti-correlated between spots in the right upper corner (red in MFEA, EMCA) and left lower corner (red in WCEU, ITAP) of the map, which reflects antagonistic switching of the mean SNP-scores of the respective spots (Figure 2D, left map, red lines). About 1000 SNPs (about 10% of all SNPs tested) were collected in these two major clusters. SNP-scores between WCEU and ITAP one hand and between a series of spots characteristic for MFEA and EMCA, on the other hand, were mutually correlated, indicating closer similarity (blue lines). Spot implication analysis complements this picture by identifying spots appearing together in the individual portraits of the cultivars. Implications are visualized in the map by connecting such joint spot appearance by lines. Again, we find three major patterns, namely co-occurrence of spots in the left, the upper and the right part of the map, respectively. Interestingly, implication and correlations links along the lower border of the map are virtually absent, which means that spots from WCEU and IBER virtually don't co-occur and are not correlated. This lack of links suggests a barrier for genomic exchange between these regions. In summary, SOM portraits visualize the genomic diversity of grapevine cultivars with single-accession and regional resolution. Overall, we identified the characteristic differences of genomic features (SNP-profiles, variance, implications and correlations) between WCEU and ITAP, IBER and cultivars from more eastern regions ranging from BALK to EMCA, MFEA and NEWO.

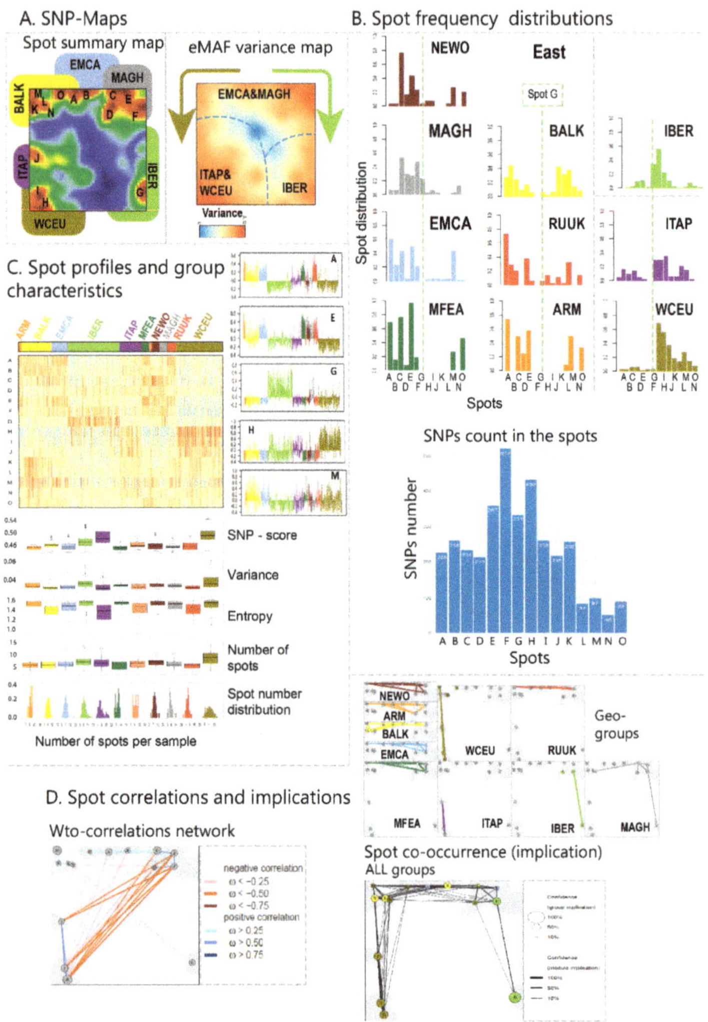

Figure 2. SNP-spot characteristics: (**A**) The SNP-maps provide information about SNP-spots (spot summary map) and variance of the allele-score (variance map). The maps roughly divide into three main areas, accumulating cultivars from WCEU, IBER and the 'East' (i.e., BALK, RUUK, MFEA, EMCA, MAGH). (**B**) Spot frequency distributions of different geographic regions assign region-specific spots, namely around G (vertical dashed line) for IBER, to the right from G (spots H–K) for WCEU and ITAP and to the left (spots A–E) for vines from the East. BALK and RUUK vines show mixed properties of WCEU and East vines. The number of SNPs per spot is largest in spots E–H. (**C**) SNP-score profiles of the spots and their geographic group-wise characteristics (mean eMAF, variance, entropy, mean spot number and spot number distributions). (**D**) Correlations between spot profiles and co-occurrence (implication) of spots. Correlations were calculated using the weighted topology-overlap (WTO) algorithm [27]. Implications were shown for all groups and separately for each of the geographic groups. The lines link spots which frequently co-occur and thus imply each other due to associations between the SNPs covered by them [13].

3.3. SNP Distributions and Associations with Vine Utilization

Next, we analyzed the distribution of selected SNPs across the map. Laucou et al. [11] identified 14 SNPs that allow one to assign the 783 cultivars studied. These SNPs are found mostly in areas of our map, which are attributed to red positive eMAF regions observed for WCEU/ITAP and IBER geographic regions (Figure S1A in Additional File 4: Supplementary Figures). Hence, cultivars were identified using SNP-markers, showing increased eMAF values in European cultivars and decreased values in Eastern ones. Interestingly, the number of these SNP-markers roughly agrees with the number of 'spot-modules' of correlated SNPs found in our SOM-analysis. Hence, our data-driven dimension reduction of feature space confirms previous results, namely, that about one-two dozen single SNPs's are sufficient to describe roughly the genetic diversity of vine cultivars [11]. Spot clusters contain typically a few hundreds of correlated SNPs that make them robust fingerprint markers of cultivar diversity. Moreover, our SNP portrayals clearly show that more subtle and more diverse SNP patterns are hidden beyond this first, relatively coarse level of genetic diversity. In principle, each accession is genetically unique with possible impacts for diverse phenotypic traits and associated genetic patterns. Their discovery requires more extended genetic characterization, e.g., by means of sequencing approaches, and more detailed genomewide association data. Our SNP-portrayal provides an option to deal with such increased data in future applications.

The chromosome-wise distribution of SNPs shows local enrichment in most cases (see the distribution of SNPs from Chr. 1 and 2 in Figure S1B in Additional File 4 as examples), however, overall there is no clear relation between the distribution of SNPs along the genome and their location in the map. Mapping of SNPs from the different spots along the chromosomes shows a similar result, namely that overall there is no clear one-to-one relation between distribution along the chromosomes and across the map, despite the local accumulation of spot-SNPs on different chromosomes (examples for six spots are shown in Figure S1C, Additional File 4).

Phenotype maps visualize the correlation between the SNP-scores and the utilization for the table, wine or double usage (Figure 3A). We find strong correlations (Pearson's correlation coefficients r = 0.7) for table usage in the right upper corner of the maps, while wine usage correlates with SNPs enriched in the lower-left corner. Correlation with double usage cultivars is smaller (r = 0.3) and refers to SNPs accumulation around spot D. The top hundred SNPs with the largest correlation coefficients localize in narrow areas in the genetic landscape. They also accumulate in different chromosomal regions (Figure 3A).

The phenotype map reflects the preferential utilization of cultivars used for table consumption in the East, while wine utilization dominates in Europe (WCEU, ITAP, BALK, IBER). Overall, we find an association between vine utilization and specific areas of the SOM, which reflects an East-to-West gradient with preferential table usage in the East and wine usage in the West in agreement with [8]. Note that these two spot-areas of the table- and wine-vine enrichment constitute the major 'axis' of anticorrelated SNP-patterns (compare with correlation net in Figure 2D), which, in turn, refers to the east-to-west 'gradient' of vine utilization established previously [10]. Interestingly, grapes for double wine and table usage refer to a region of the map linking MAGH and IBER cultivars, i.e., a region merging Islamic and Christian traditions (see [10] and Discussion). Another 'transition' region are BALK and ITAP, where however wine and table grapes are used in parallel. Additionally, the frequency distributions of cultivars, according to their utilization, supports the East-West gradient (Figure 4C). The table utilization of grapes dominates in the east and in MAGH while wine utilization dominates in Europe, where the region between MAGH and IBER forms a transition range enriching grape of double wine and table usage.

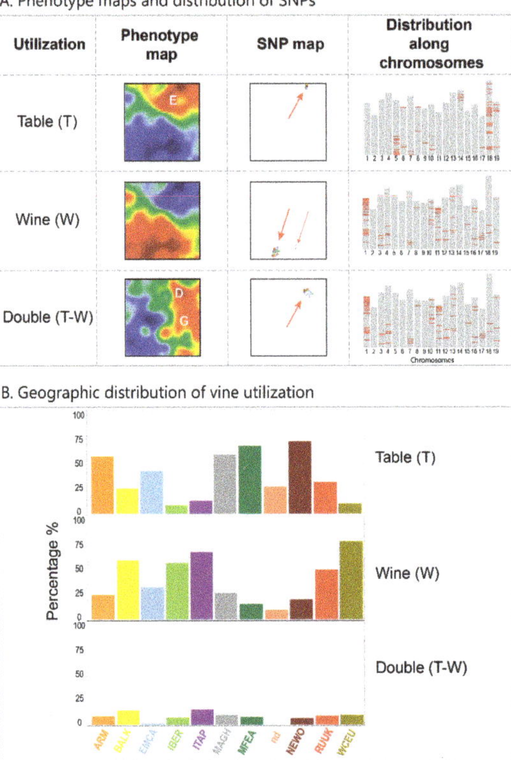

Figure 3. Association with vine utilization: (**A**) Phenotype maps color code correlation between SNP-scores and cultivar-utilization (table, wine, double) between red (high correlation) and blue (low correlation). Top 100 correlated SNPs accumulate in narrow regions of the SNP map (see arrows) in and around spots E (table utilization), H (wine) and D (double) and on selected chromosomes (assembly version 12X.V2). (**B**) Distribution of vine utilization (percentage of the respective vine cultivars) indicates enrichment of table utilization (T) in Eastern cultivars, of wine utilization (W) in western ones (WCEU, ITAP, IBER, BALK), while the distribution of double usage cultivars partly resembles that of wine vines however with increased frequencies in MFEA and MAGH.

A. Metronet' of country-wise ANOVA-portraits

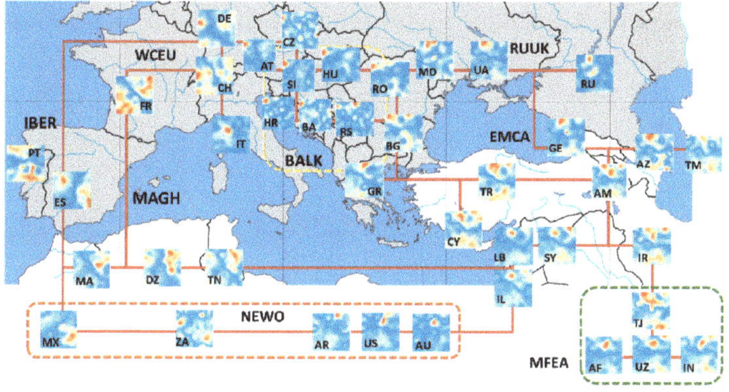

B. Dissemination paths of grapes in sample and SNP-space

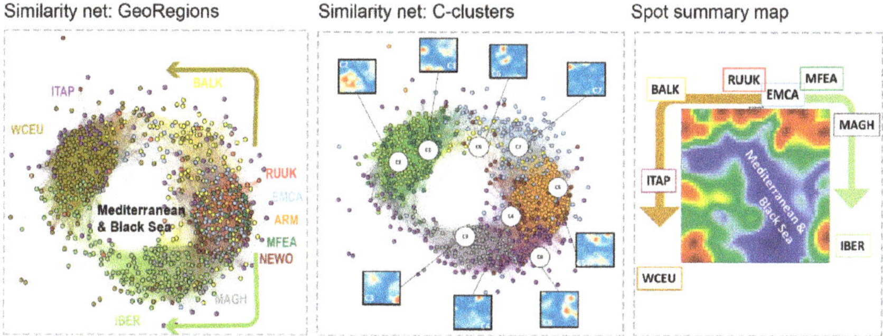

Figure 4. Distribution of grapevines around the Mediterranean and Black Sea: (**A**) Country-wise ANOVA portraits of cultivars are linked by a 'metro-net', which visualizes the relatedness between similar portraits in neighboring countries. (**B**) The network of mutual correlations between the SNP-portraits arranges them in a ring-shaped fashion, resembling the ordering of geographic regions around the Mediterranean and Black Sea. Moreover, the activated spots in the SOM order roughly in this way. Recoloring of the cultivars using the eight 'C'-clusters taken from [11] sorts them in a consecutive way.

3.4. Genetic Relatedness of Cultivars around the Mediterranean Sea

To increase geographic resolution, we calculated country-wise SNP-landscapes using ANOVA (analysis of variance)-portrayal (Figure 4A). The ANOVA-portraits show regions of the large variance of the SNP-score in red and of low variance in blue. Inspection of the country-related portraits reveals overall a high diversity, where, however, portraits of countries from the same region often resemble each other. Moreover, the textures alter in a systematic way between the regions, e.g., from the east (EMCA, MFEA) to the west (via RUUK, BALK to WCEU, ITAP and IBER), as visualized by the 'metro-net' lines linking similar country portraits. Accordingly, one finds similarities of these portraits for neighboring countries from Georgia, via Russia, Ukraine and Moldova, towards Balkan countries into the west direction and from Georgia and Armenia via Iran towards Tadjikistan, Uzbekistan and Afghanistan, into the Middle Asia region (MFEA). Another route is directed from the Caucasus via Lebanon, Israel towards North Africa (MAGH) and Iberian Peninsula (IBER). The South Caucasus is also linked via Anatolia (Turkey), Cyprus with the Balkan. In the western part of Europe, portraits from Spain show similarities with Northern African countries (MAGH), and only partly with French and

German portraits, which, in turn, show similarity links via Switzerland, Austria and the Czech Republic towards Balkan. Mexican cultivars resemble Spanish ones according to their SNP-portraits while cultivars from USA, Australia and Argentina, on average, reflect more similarities with grapevines from MFEA and EMCA.

In order to evaluate the relatedness between the individual SNP-portraits of cultivars, we calculated similarity nets. They are based on Pearson's-correlation coefficients between the individual SOM-portraits and arrange them in a way that highly correlated cultivars were connected by lines and locate in the close mutual distance (Figure 4B, left). The net divides into the three major clusters enriched in cultivars from WCEU (and ITAP), MFEA and IBER, respectively, which were separated by regions of decreased accession density, especially between the WCEU and IBER clusters, reflecting reduced numbers of correlation links. The spot summary map (Figure 4B, right part) indicates exactly the same properties in SNP-feature space: the blue, low eMAF region separates spots of high eMAF-values observed in WCEU, IBER and the 'East'. Importantly, these regions group around the Mediterranean and Black Seas, suggesting more close genetic relatedness among the cultivars along a northern and a southern coast-route, respectively (see arrows in Figure 4B). Besides the clear enrichment of cultivars from different geographic regions in the clusters, one also finds the considerable intermixing of them, reflecting genetic exchanges between the regions (see next subsection). The stratification of accessions into eight clusters (C1–C8), as proposed previously [11], provides a better 'resolution of the correlation structure of cultivars in the similarity net (Figure 4B, C-clusters). Hence, 'country-wise' SNP-(ANOVA)-portrayal provides an intermediate level of resolution in-between the geographic regions and the individual cultivar portraits. It enabled us to establish their mutual relatedness, and to identify similarity links between them, ranging from the Middle East and the Caucasus to Iberia. Main roads of relatedness between the SNP-portraits of grapevine cultivars are found along a northern and a southern route about the Mediterranean and the Black Sea. A marked genetic gap is found between IBER and WCEU, suggesting that the Pyrenees are the main genetic barrier in the West. 'Eastern' cultivars, especially from the MFEA and EMCA regions, form a relatively tight cluster of mutually similar genomes. It links to the western cultivars via BALK and MAGH, along the northern and southern routes, respectively.

3.5. Diversity Trees Suggest South Caucasus as a Crossroad of Grapevine Dissemination

In the next step, we analyzed SNP-portraits and their diversity on the level of individual cultivars. An inspection of their gallery (Additional File 1) indicates typical patterns for their geographic regions, which, however, partly mix between them. Hierarchical clustering trees of cultivar portraits selected from four different regions illustrate this mixing effect (Figure 5A). For example, cultivars from RUUK split into groups resembling genomic patterns from EMCA, BALK and WCEU, respectively. Cultivars from MAGH divide into groups resembling IBER and ITAP patterns, but also a larger group of 'local', MAGH-like ones. To quantify this intermixing, we applied re-distribution analysis (Figure 5B), which uses a cluster stability score to estimate whether a certain accession better fits into its geographic region (positive score value) or into another one (negative score value). MFEA cultivars show high intra-cluster stability, followed by WCEU, BALK and MAGH, while, e.g., RUUK cultivars tend to distribute virtually completely into other clusters. Hence, the MFEA region can be interpreted as a sink for cultivars from other geographic regions, except for ITAP and, partly, WCEU. EMCA genomes are similar to MEFA genomes, as indicated by the fact that the latter ones (including ARM) nearly exclusively tend to shift into the cluster of the former one.

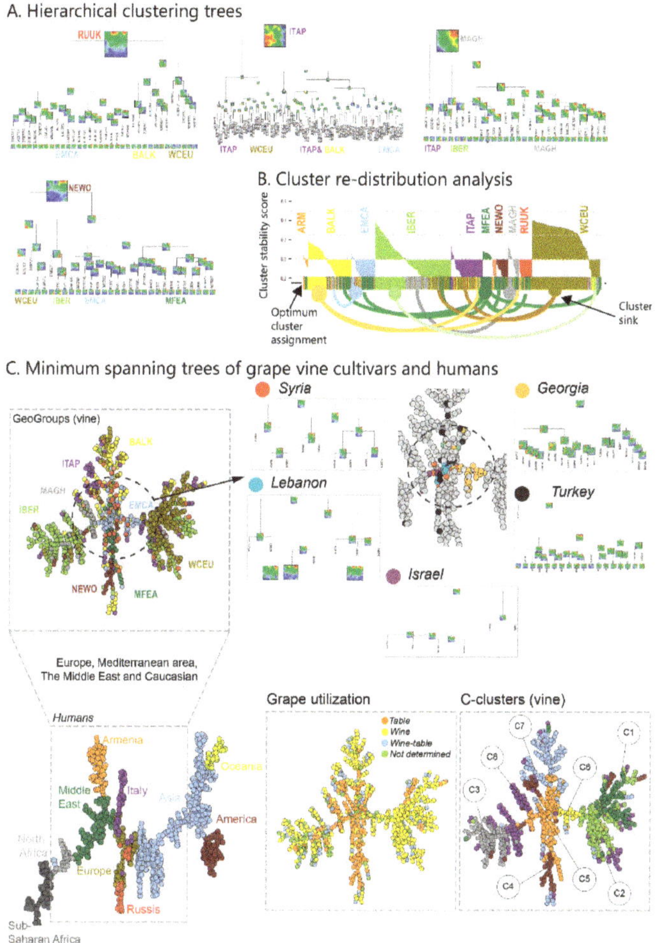

Figure 5. Genome diversity in and between the geographic regions: (**A**) Hierarchical clustering of vine cultivars from selected regions shows overlapping genomic properties. (**B**) The cluster stability plot ('silhouette plot') indicates intra- and inter-cluster similarities. The similarity score is positive for stable cluster-assignment of cultivars and negative, if a cultivar better matches into another cluster (the best-matching cluster for each cultivar is indicated by the colored bar below the silhouette plot). The arc-lines link unfavorable cluster assignments with the respective sink (best matching)-cluster. (**C**) The minimum spanning tree (MST) of the vine cultivars reveals four major leafs enriching IBER (and MAGH), BALK (and ITAP), WCEUR and MFEAS (and NEWO) grapes, respectively. The central 'cross-road' linking the four leafs enriches EMCA-cultivars. Particularly, cultivars from Lebanon, Israel and Syria locate within the 'cross-road' region (see an enlargement in the right part). Vines from Georgia form the trunk of the European part. Armenian and Turkish vines spread more widely along the MST. Re-coloring the MST according to grape utilization shows that table usage predominates along the vertical trunk of the MST (MCEU, EMCA) and along the MAGH-branch, while wine usage is found mainly for European grapes. Alternative clustering, as proposed previously (C-clusters) [11], better separates the clusters along the MST. The MST of human genomes of modern populations from Africa to Oceania taken from the Human Genome Project (HGP) shows a more linear arrangement (left part), reflecting different dissemination patterns. The dashed rectangle includes populations of the same geographic regions, as studied for the grapevine cultivars.

To further evaluate the relatedness between the cultivars, we calculated a minimum spanning tree (MST), which, in contrast to the correlation net discussed above, connects only most similar cultivars and, in consequence, generates 'paths' of closest similarities (Figure 5C). The MST obtained reveals a cross-like shape, showing four major branches. Cultivars from IBER (left branch), WCEU (right branch), BALK and ITAP (top) and MEFA (down) accumulate in the four major branches, while EMCA cultivars are found predominantly in the central crossroad area (Figure 5C). The 'cross-like' structure of grapevine cultivars suggests their distribution along four paths with their common origin in the EMCA area. Zooming in reveals that Georgian cultivars form the trunk of the WCEU-branch to the right, while cultivars from Syria, Lebanon and Israel are at the origin of the trunk to the left IBER and MAGH branch. Notably, Georgian accessions split in portraits more similar to western Europeans, to Greek ones and to typical EMCA portraits (see the hierarchical clustering tree of Georgian vines in Figure 5C), which reflects a large variability of their genomes. Armenian and Turkish cultivars distribute more in the vertical direction along the MFEA and BALK branches. Coloring of the MST according to grape utilization shows strong enrichment of table cultivars along the vertical trunk enriching MFEA grapes, while wine cultivars enrich in the European vine branches referring to WCEU, IBER and BALK. Alternative recoloring the MST using the alternative 'C'-clustering [11] better adjusts to the cross-like structure, where C5 collects the cultivars from the crossroad area, except the Georgian ones (C6) (Figure 5C). Interestingly, WCEU cultivars split into a C1 and C2 clusters with C1 collecting cultivars from Germany and Austria, while C2 enriches cultivars from Italy.

Note, that the structure of the vine-MST is completely different from the MST of human genomes of worldwide populations taken from the human genome project. This MST of human genomes possesses an almost linear backbone, with only a few side branches. It reflects the 'out of Africa' basic migration history of modern humans, which initially proceeded mostly along one major path. Hence, MST presentation seems to reflect the mechanisms and paths of genome dissemination. In summary, diversity analysis on single cultivar level reveals a marked intermixing between the different geographic regions where, however, cultivars from the South-Caucasus and Near East from the trunk for diverging branches of vines are typical for IBER (and ITAP), WCEU, MFEA and BALK (and RUUK) regions, respectively.

3.6. 'Pseudotime' Suggests Different Scales of Cultivar Diversification

For a more detailed study of possible paths of the distribution of grapevine, we applied so-called 'pseudotime' (PT) analysis. It arranges cultivars according to their mutual genetic similarity along possible multi-branched paths, using a diffusion-law based algorithm [26]. The obtained paths thus describe the dissemination of grapevine cultivars between the geographic regions, assuming a starting area which we set to cultivars from the 'cross-road' part of the MST (see the previous section). PT-analysis assigns a similarity measure called 'pseudotime', normalized between zero (start) and unity (end). The grapevine accessions sort into four branches, where each of them collects cultivars of mixed geographic origin (Figure 6A, left part). At the root the tree divides into two branches (numbered '0' and '4'), where '0' collects predominantly MFEA, BALK and ITAP cultivars and '4' enriches grapes from IBER and MAGH. Branch '0' further splits at larger PTs into three sub-branches ('0–1' to '0–3') accumulating WCEU, ITAP and again IBER vines, respectively. Mean SOM-portraits along the branches reveal clear differences in the SNP-landscapes, supporting their assignment to BALK and RUUK (branch 0), WCEU and ITAP (0–1), WCEU (0–2), IBER (0–3) and MFEA and EMCA (4) characteristics (Figure 6A, middle column). Moreover, t-SNE clustering provides a more diverse stratification of cultivars into ten groups, where most of them were characterized by one-to-two major spots in their portraits (Figure 6A, right part). Importantly, spots appear in a consecutive fashion along the paths; this way providing developmental traces in the SNP-landscape as illustrated by the white curves in the SOM-overview map (Figure 6A, right part).

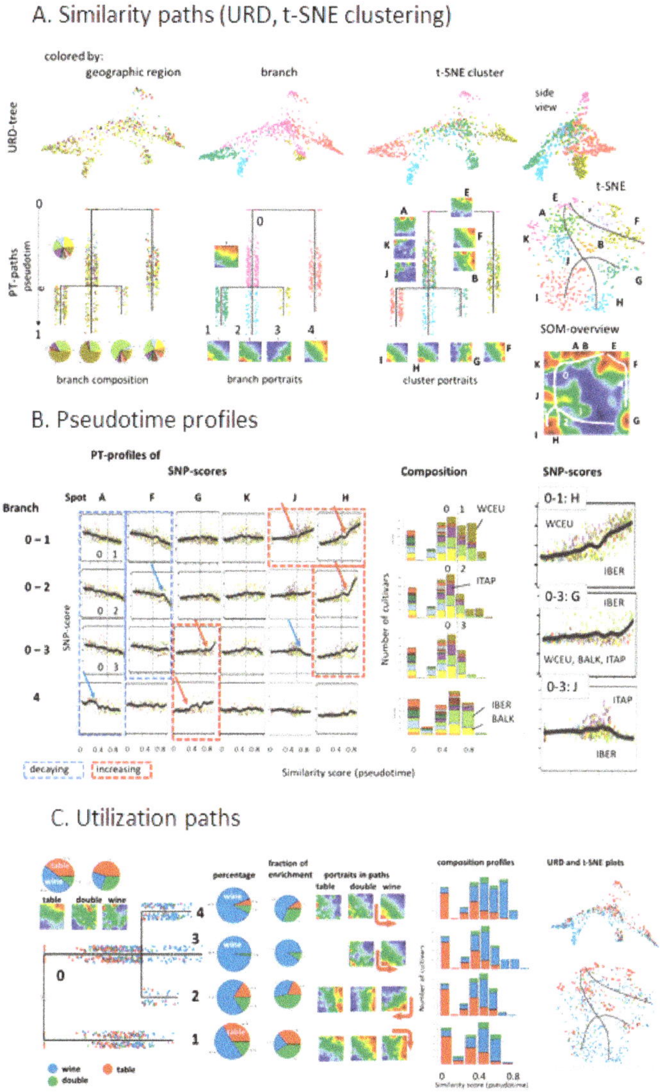

Figure 6. Multibranched pseudotime analysis: (**A**) Similarity paths using URD [26] reveal a 'Kraken' like distribution of vine cultivars which is described by four 'tentacles' as pseudotime branches. They accumulate cultivars preferentially from WCEU and IBER (see pie diagrams), respectively. t-SNE clustering provides ten clusters labelled by the most prominent SNP-spot expressed in each of them (right part). The grey and white curves visualize the paths of cultivar similarities in the t-SNE accession- and SOM SNP-space, respectively. (**B**) Pseudotime analysis reveals increasing (marked by red dashed rectangles) and decaying (blue rectangles) courses, indicating, e.g., accumulation or depletion of WCEU cultivar features and accessions at later PT values. Arrows indicate points of increasing (red) and decreasing (blue) slopes. Composition of the geographic origin of the cultivars is shown as barplot (middle). Enlarged SNP-profiles, as shown in the right part of the figure, reveal details of cultivar distribution around the LOESS-fits (locally weighted scatterplot smoothing, black curves).

(**C**) The utilization of grapes (wine, table, double wine and table) along the developmental paths indicates that wine usage appeared at later PT and along all branches compared with table utilization. The pie diagrams show percentages of grape utilization along the branches (larger diagrams) and relative percentages (smaller pie diagrams) as ratio percentage along the branch divided by the overall percentage (wine: 61%, table: 25%, double: 12%). The mean SOM-portraits refer to cultivars of different utilization along the branches. SNP-spots shift into the direction as indicated by the arrows.

The consecutive alterations in feature space observed were supported by PT-profiles of selected spots (Figure 6B). The eMAF SNP-score of spot A (up in EMCA) decays along all branches (blue, dashed frames), reflecting increasing genetic distance from this region with increasing PT. In contrast, spot F (up in MFEA and MAGH) remains relatively invariant along branch 4. The IBER and WCEU related spots gain in 'later' PT along different paths (red dashed frames), namely, path 4, 0–1 and 0–2, respectively. These courses, however, differ in their turning points where eMAF starts to increase and in their slopes. For example, the gain of 0–2 starts at later PTs and it is steeper compared with 0–1, suggesting different 'switching points' and time-scales of grapevine distribution (see red arrows). Interestingly, ITAP-cultivars accumulate at intermediate PTs along the spot profile J of all subbranches 0–1, –2 and –3. Bar plots of the composition of cultivars along the branches, as a function of PT, support the accumulation of MFEA, EMCA, RUUK, MAGH, and partly ITAP vines at early and intermediate PTs while WCEU and IBER (and partly BALK along path 4) cultivars accumulate at later PTs. Hence, PT-scaling obviously quantifies aspects of dissemination of cultivars in East-to-West direction via different routes.

The consideration of grape utilization along the different PT-paths clearly reveals the enrichment of table vines in branches 0 and 4, while wine cultivars strongly enrich in the branches 1–3 at later PTs (Figure 6C, pie diagrams). The gain of wine usage also becomes evident in the PT-profiles at PT > 0.5, and, on the other hand, table vine enrichment at PT < 0.5 (Figure 6C, bar plot). Stratification of the portraits along the paths reflects the systematic shift of spot patterns between the portraits of the table and wine vines from the upper to the lower parts of the maps along the paths shown in Figure 6A (see also the arrows in Figure 6C, middle). The URD- and t-SNR plots support this result by the separation between table (red color) and wine (blue) cultivars in top-to-down direction (right part in Figure 6C).

In summary, multibranched PT-analysis resolves the 'transverse' distribution of cultivars among geographic regions also in 'longitudinal' PT-dimension, by sorting them according to mutual similarities of their genomic landscapes. We find the accumulation of grapes used for wine production at later PTs especially in Europe, which suggests the diversification of wine cultivars at later times compared with table grapes.

4. Discussion

We applied SOM artificial neural network-based machine learning to genome-wide SNP data of 783 grapevine cultivars, in order to visualize and to analyze the landscape of grapevine genomes, in terms of topological features such as 'mountains' and 'valleys', referring to positive and negative eMAF values, respectively (Figure 7).

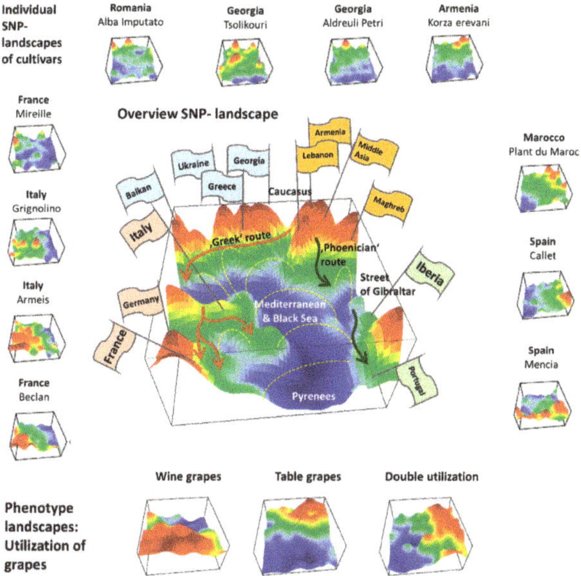

Figure 7. Genomic landscapes of grapevine cultivars: The SNP-landscapes of all 783 cultivars (examples from different regions are shown as small images) were summarized into a mean 'overview' landscape (large image with flags). Correlated SNP-profile self-organized together due to Figure 7 examples are shown as small images). The 'portraits' project the multidimensional similarity space spanned by the excess minor allele frequencies (eMAF) of the 10,207 SNPs under study into two dimensions. The vertical 'height' dimension visualizes the eMAF-value of the SNPs defined as the deviation of its minor allele frequency (MAF) in a certain cultivar from its mean MAF-value averaged over all cultivars. The algorithm 'self-organizes' similar SNP-profiles more closely together than different ones. In the result, we obtained 783 different SNP landscapes, each showing a unique combination of red 'mountain-ridges' and blue 'valleys' referring to positive and negative eMAF-values, respectively. The overview landscape summarizes the diversity of the individual portraits in terms of frequently observed SNP-patterns (Figure 7, large image). Its topology reveals several 'mountain areas' which refer to cultivars from different geographic regions. It relates to the geographic origin of grape cultivars, their phenotypes and history of distribution throughout the 'classical' world, starting from its origin of domestication about 10,000 years ago. Particularly, the genomic landscape resembles the geographic map around the Mediterranean and the Black Sea. Namely, the eMAF-'mountains' order cultivars from the Caucasus along a northern route' via Balkan towards Western Europe and along a southern route via Palestine and Maghreb towards the Iberian Peninsula. A central 'blue valley' referring to predominantly negative eMAF-values separates both routes. It can be interpreted geographically as the Mediterranean and Black Sea areas, which obviously constitute areas of reduced genetic exchange. Interestingly, the largest barrier is found between grapes from the Iberian Peninsula and Western Europe (France, Italy), while the street of Gibraltar appears only as a small sidearm of the central 'genetic' valley, thus indicating a relatively moderate genetic barrier between North Africa and Iberia. Hence, Iberian Peninsula and northern Africa could be considered subcontinents of Europe, being separated by the Pyrenees according to vine genetics, which also agrees with the recent comparison of human genomes from these areas [28]. Another moderate genetic barrier is found between grapes from the Balkan and Western Europe (Germany, Switzerland and Italy). According to these barriers, cultivars divide into three major groups on the coarsest level of classification, namely Western Europe and Italian grapes, Iberian grapes and vine cultivars from Eastern and Maghreb regions. Detailed inspection of the mountain range of 'eastern' grapes reveals fine internal structure of valleys separating, e.g., Armenian from Georgian grapes and vines from Anatolia and Greece from Balkan ones.

We complemented this genetic feature (SNP-) landscape by grapevine accession-similarity plots. Each of them visualizes different aspects of the mutual relatedness of cultivars. Principal component plots distribute accessions in (eMAF) distance-scale. It results in local overcrowding of cultivars with similar genomes, e.g., from Near East, Caucasus and Middle Asia while Western European and Iberian cultivars separate into virtually different data clouds of large variability between sample genomes. PCA requires separate projection-plots to cover the essential properties of similarity space. The (correlation) similarity net projects multidimensional sample space into two dimensions. It reveals major routes of grape distribution around the Mediterranean Sea and the genetic barrier between Iberian and Western European cultivars, evident also in the feature landscape. In contrast, the minimum spanning tree (MST) generates path-like structures. For worldwide human genomes, the MST reflects roughly the 'out-of-Africa' migration history, in terms of a virtually linear path starting from Sub-Saharan Africa and ending in East Asia, with side branches towards Europe, America and Oceania. Interestingly, the grapevine MST splits into four mean branches, enriching West European, Iberian, Balkan and Middle East grapes, and with cultivars from South Caucasus and Fertile Crescent (Syria, Lebanon, East Anatolia) at their central crossroad. A detailed comparison of individual (ANOVA-) portraits further supports this result, using the feature landscapes in a country-wise resolution.

Overall, our genomic landscape and the different sample similarity plots are consistent with the historical knowledge and previous genetic findings of grapevine domestication and distribution [7,10]. Accordingly, cultivated grapes occurred initially in South Caucasus (Georgia, Armenia) and Fertile Crescent (East Anatolia and North Lebanon and Syria) [29,30]. Indeed, grapes from this region form the crossroad in the MST, presumably due to footprints of initial cultivation in their genomes. Grapes then distributed towards the 'classical', Mediterranean world into the west direction and into East towards Iran and the Middle East (Tadjikistan, Uzbekistan), Afghanistan and India. The northern and southern ways into the West agree with the distribution of settlements of Greeks (the Black Sea, including Crimea, Anatolia, Southern Italy and Sicily, Southern France and Northern Spain along the Northern coast of Mediterranean Sea) and Phoenicians (Lebanon, Carthago/Tunis, Maghreb and South-West Spain), respectively. Genomes of vines from the Italian Peninsula are very diverse and reflect links to almost all parts of the Mediterranean area (BALK, EMCA, MAGH, IBER) and Western Europe, presumably due to intense cultural exchange in the Greek world, and later within the Roman empire [31].

Hence, grape growing and winemaking in classical time disseminated from 'Iberia-to-Iberia', starting in or near the ancient kingdom Iberia in Middle Georgia in South Caucasus in the East and ending at the Iberian Peninsula in the west. The combined action of selection, breeding, admixture and migration have shaped the cultivated grape diversity. Substantial genetic diversity has been maintained, subsequent to domestication derived from Transcaucasian Wild Grape, possibly due to several events of introgression from local wild vine Silvestris-varieties, Wild grapes (*V. vinifera* ssp. sylvestris), particularly in Western Europe [7]. Although Mediterranean and Black Seas served rather as highways of cultural exchange than as barriers, vine distribution obviously followed primarily 'country ways' along with the coastal areas.

We visualize grape utilization in terms of phenotype maps which associate table, wine and double usage with different geographic regions (Figure 7, part below). Grapes for fresh consumption (table vines) predominate in Eastern and North African areas, while wine utilization is found mostly in Western Europe (RUUK, BALK, WCEU, ITAP, IBER). This division has been associated with different religious rules concerning alcohol consumption in Islamic and Christian regions [8]. Interestingly, the red region of wine-utilization covers the WCEU and IBER geographic regions, and thus it bridges the genetic border formed by the Pyrenees in the overview landscape (Figure 7).

In order to extract additional information about the dissemination of cultivars, we applied pseudo-time (PT) analysis. This method has been originally developed for describing cell differentiation using single-cell transcriptomic data [26,32], and it was applied to study human cancer progression using single-cell and bulk transcriptome data [14,33]. In the context of this study, PT sorts and scales vine genomes according to their genetic similarity, into a directed, branched graph. Starting with the assumption of initial Eastern domestication of the grapevine PT then describes its dissemination in an East to West direction. Vines from the Caucasus and the East refer to small PT-values and vines from the Balkan and Italian peninsula mostly to intermediate PTs. Interestingly, the genetic characteristics of Western European and Iberian grapes follow different courses at later PTs along different branches, which eventually reflects different waves of cultivation, since the Roman Empire times. Particularly, recent pedigree analyses uncover a putative Middle Age cultivar melting pot, giving rise to many of today's cultivars, suggesting even secondary domestication events taking place in Western Europe and the Iberian Peninsula ending in the cultivars that are used in viticulture today [8]. Different PT-branches show different slopes and switching points in their grow regimes at later PTs. Table grapes mostly accumulate at early PTs. Hence, our 'SOMelier' transforms vine genomes into a landscape resembling the topology of the geographic regions of grape cultivation and/or the diversity of their genomes. Different methods of similarity analyses in SNP-feature and cultivar space enable extracting details of dissemination history and the utilization of grapes.

5. Conclusions

Our 'SOMmelier' approach visualizes essential aspects of grapevine genomes related to geographic distribution, paths of dissemination and vine utilization. 'SOMmelier' portrays vine genomes in terms of individualized images. They are intuitive, meaning that they don't need specialized genetic knowledge for interpretation. Together with vine genome landscapes described here, we propose to use individual genomic portraits as an option to supplement vine cultivar passports as fingerprint characteristics of their genomes. Such fingerprint portraits consider virtually the whole diversity of the vine. Note also that we here discuss accession portraits mainly on the level of geographic regions as a sort of worked example to support the interpretation of genomes of individual accession portraits. Our study also demonstrates that bioinformatics methods proven before in analytic tasks on different omics realms, mostly transcriptomics, but also epigenomics, proteomics and human genomics, provide reasonable results if applied to vine cultivars as an example of plant genomes. Our approach thus extends the methods toolbox for plant genetics by providing novel approaches which complement established ones. Their pros and cons should be evaluated in future applications. The 'SOMelier 'method opens the opportunity to process larger genotype data, obtained by, e.g., whole genome sequencing and/or increased number of cultivars included.

Supplementary Materials: The following are available online at http://www.mdpi.com/2073-4425/11/7/817/s1 Additional File 1: Individual cultivar accession portraits (PDF-format). Samples are assigned according to sample IDs taken from Loucou et al. [11]. Additional File 2: The list of SNPs collected in all spots. Each sheet includes the list of SNPs for all SOM spots (A - O) (XLS-format). Additional File 3: Summary annotation of SNPs, used in this study [11] (CNV-format). Additional File 4: Additional figures Figure S1 and Figure S2 (PDF-format): Figure S1: SNP maps, eMAF profiles and associations with selected phenotypes. (a) 14 SNPs (assembly version 12X.V0) selected in [11] to identify the 783 vine cultivars accumulate in WCEU/ITAP and IBER regions of the map. Their mean SNP profile shows large eMAF-values for cultivars from these regions on the average. Profiles of selected individual SNPs reveal a nearly bi-modal distribution of eMAF-values. (b) SNPs from different chromosomes distribute over several spot regions (red frames). (c) Distribution of SNPs from selected spots (red marks) among the 19 chromosomes (grey vertical bars). Figure S2: ANOVA-portraits stratified according to geographical regions and utilization of grapes. The log P-value is visualized on the SOM map. The most significant areas for all phenotypes overlap with characteristic spots for MAGH cultivars.

Author Contributions: Conceived and lead study, H.B. and A.A.; Bioinformatics analyses, M.N., M.S.; Drafting manuscript, H.B., M.N.; Methods support: H.L.-W.; Vine cultivar expertise: K.M.; All authors read and approved the final version of the manuscript.

Funding: This study was supported by the German Federal Ministry of Education and Science (BMBF) grants LHA (idSEM program: FKZ 031L0026 to H.B. and HL.-W.), PathwayMaps (WTZ ARM II-010 and 01ZX1304A to H.B. and A.A.), and oBIG (FFE-0034 to HL-W), DAAD Short-Term Grant—2019 (57440917). A.A. and M.N. were supported by the Internal Grant of Russian-Armenian university within the framework of funding from the Ministry of Education and Science of the Russian Federation.

Conflicts of Interest: All authors declare no conflict of interest.

Abbreviations

Abbreviations of counties

NEWO	New World
MAGH	Maghreb
EMCA	Eastern Mediterranean and Caucasus
MFEA	Middle and Far East
BALK	Balkans
RUUK	Russia and Ukraine
ARM	Armenia
IBER	Iberian Peninsula
ITAP	Italian Peninsula
WCEU	Western and Central Europe

References

1. Zohary, D. The domestication of the grapevine *Vitis Vinifera* L. in the near east. In *The Origins and Ancient History of Wine: Food and Nutrition in History and Antropology*; McGovern, P., Fleming, S., Katz, S., Eds.; Routledge: London, UK, 2003. [CrossRef]
2. Olmo, H.P.; Simmonds, N.W. *Evolution of Crop Plants*; Longman: London, UK, 1976; pp. 294–298.
3. Rusishvili, N. *The Grape Vine Culture in Georgia on the Basis of Palaeobotanical Dat*; Mteny Association: Tbilisi, Georgia, 2010; Volume 3.
4. Riaz, S.; De Lorenzis, G.; Velasco, D.; Koehmstedt, A.; Maghradze, D.; Bobokashvili, Z.; Musayev, M.; Zdunic, G.; Laucou, V.; Walker, M.A.; et al. Genetic diversity analysis of cultivated and wild grapevine (Vitis vinifera L.) accessions around the Mediterranean basin and Central Asia. *BMC Plant Biol.* **2018**, *18*, 137. [CrossRef] [PubMed]
5. Henning, F.-W. *Wine and the Vine. An Historical Geography of Viticulture and the Wine Trade*; Routledge: London, UK, 1992; p. 430.
6. Negrul, A.M. Evolucija kuljturnyx form vinograda. *Doklady Akademii nauk SSSR* **1938**, *8*, 585.
7. Myles, S.; Boyko, A.R.; Owens, C.L.; Brown, P.J.; Grassi, F.; Aradhya, M.K.; Prins, B.; Reynolds, A.; Chia, J.-M.; Ware, D.; et al. Genetic structure and domestication history of the grape. *Proc. Natl. Acad. Sci. USA* **2011**, *108*, 3530–3535. [CrossRef] [PubMed]
8. Migicovsky, Z.; Sawler, J.; Gardner, K.M.; Aradhya, M.K.; Prins, B.H.; Schwaninger, H.R.; Bustamante, C.D.; Buckler, E.S.; Zhong, G.-Y.; Brown, P.J.; et al. Patterns of genomic and phenomic diversity in wine and table grapes. *Hortic. Res.* **2017**, *4*, 17035. [CrossRef]
9. Liang, Z.; Duan, S.; Sheng, J.; Zhu, S.; Ni, X.; Shao, J.; Liu, C.; Nick, P.; Du, F.; Fan, P.; et al. Whole-genome resequencing of 472 Vitis accessions for grapevine diversity and demographic history analyses. *Nat. Commun.* **2019**, *10*, 1190. [CrossRef]
10. Bacilieri, R.; Lacombe, T.; Le Cunff, L.; Di Vecchi-Staraz, M.; Laucou, V.; Genna, B.; Péros, J.-P.P.; This, P.; Boursiquot, J.-M. Genetic structure in cultivated grapevines is linked to geography and human selection. *BMC Plant Biol.* **2013**, *13*, 25. [CrossRef]
11. Laucou, V.; Launay, A.; Bacilieri, R.; Lacombe, T.; Adam-Blondon, A.F.; Bérard, A.; Chauveau, A.; Andrés, M.T.d.; Hausmann, L.; Ibáñez, J.; et al. Extended diversity analysis of cultivated grapevine Vitis vinifera with 10K genome-wide SNPs. *PLoS ONE* **2018**, *13*, e0192540. [CrossRef]
12. Binder, H.; Wirth, H. Analysis of Large-Scale OMIC Data Using Self Organizing Maps. In *Encyclopedia of Information Science and Technology*; Mehdi Khosrow-Pour, D.B.A., Ed.; IGI Global: Hershey, PA, USA, 2015; pp. 1642–1653. [CrossRef]
13. Loeffler-Wirth, H.; Kreuz, M.; Hopp, L.; Arakelyan, A.; Haake, A.; Cogliatti, S.B.; Feller, A.C.; Hansmann, M.-L.; Lenze, D.; Möller, P.; et al. A modular transcriptome map of mature B cell lymphomas. *Genome Med.* **2019**, *11*, 27. [CrossRef]
14. Kunz, M.; Löffler-Wirth, H.; Dannemann, M.; Willscher, E.; Doose, G.; Kelso, J.; Kottek, T.; Nickel, B.; Hopp, L.; Landsberg, J.; et al. RNA-seq analysis identifies different transcriptomic types and developmental trajectories of primary melanomas. *Oncogene* **2018**, *37*, 6136–6151. [CrossRef]

15. Wirth, H.; Löffler, M.; von Bergen, M.; Binder, H. Expression cartography of human tissues using self organizing maps. *BMC Bioinform.* **2011**, *12*, 306. [CrossRef]

16. Hopp, L.; Nersisyan, L.; Löffler-Wirth, H.; Arakelyan, A.; Binder, H. Epigenetic Heterogeneity of B-Cell Lymphoma: Chromatin Modifiers. *Genes* **2015**, *6*, 1076–1112. [CrossRef] [PubMed]

17. Hopp, L.; Willscher, E.; Löffler-Wirth, H.; Binder, H. Function Shapes Content: DNA-Methylation Marker Genes and their Impact for Molecular Mechanisms of Glioma. *J. Cancer Res. Updat.* **2015**, *4*, 127–148. [CrossRef]

18. Binder, H.; Wirth, H.; Arakelyan, A.; Lembcke, K.; Tiys, E.S.; Ivanisenko, V.A.; Kolchanov, N.A.; Kononikhin, A.; Popov, I.; Nikolaev, E.N.; et al. Time-course human urine proteomics in space-flight simulation experiments. *BMC Genomics* **2014**, *12*, S2. [CrossRef] [PubMed]

19. Nikoghosyan, M.; Hakobyan, S.; Hovhannisyan, A.; Loeffler-Wirth, H.; Binder, H.; Arakelyan, A. Population levels assessment of the distribution of disease-associated variants with emphasis on Armenians - A machine learning approach. *Front. Genet.* **2019**, *10*, 394. [CrossRef]

20. Hopp, L.; Löffler-Wirth, H.; Galle, J.; Binder, H. Combined SOM-portrayal of gene expression and DNA methylation landscapes disentangles modes of epigenetic regulation in glioblastoma. *Epigenomics* **2018**, *10*, 6. [CrossRef] [PubMed]

21. Wirth, H.; von Bergen, M.; Murugaiyan, J.; Rösler, U.; Stokowy, T.; Binder, H. MALDI-typing of infectious algae of the genus Prototheca using SOM portraits. *J. Microbiol. Methods* **2012**, *88*, 83–97. [CrossRef] [PubMed]

22. Malovichko, Y.V.; Shtark, O.Y.; Vasileva, E.N.; Nizhnikov, A.A.; Antonets, K.S. Transcriptomic Insights into Mechanisms of Early Seed Maturation in the Garden Pea (*Pisum sativum* L.). *Cells* **2020**, *9*, 779. [CrossRef] [PubMed]

23. Löffler-Wirth, H.; Kalcher, M.; Binder, H. OposSOM: R-package for high-dimensional portraying of genome-wide expression landscapes on bioconductor. *Bioinformatics* **2015**, *31*, 3225–3227. [CrossRef] [PubMed]

24. Bergström, A.; McCarthy, S.A.; Hui, R.; Almarri, M.A.; Ayub, Q.; Danecek, P.; Chen, Y.; Felkel, S.; Hallast, P.; Kamm, J.; et al. Insights into human genetic variation and population history from 929 diverse genomes. *Science* **2020**, *367*, 6484. [CrossRef]

25. Haber, M.; Mezzavilla, M.; Xue, Y.; Comas, D.; Gasparini, P.; Zalloua, P.; Tyler-Smith, C. Genetic evidence for an origin of the Armenians from Bronze Age mixing of multiple populations. *Eur. J. Hum. Geneti.* **2016**, *24*, 931–936. [CrossRef]

26. Farrell, J.A.; Wang, Y.; Riesenfeld, S.J.; Shekhar, K.; Regev, A.; Schier, A.F. Single-cell reconstruction of developmental trajectories during zebrafish embryogenesis. *Science* **2018**, *360*, 6392. [CrossRef] [PubMed]

27. Hopp, L.; Wirth, H.; Fasold, M.; Binder, H. Portraying the expression landscapes of cancer subtypes: A case study of glioblastoma multiforme and prostate cancer. *Syst. Biomed.* **2013**, *1*, 99–121. [CrossRef]

28. Hernández, C.L.; Pita, G.; Cavadas, B.; López, S.; Sánchez-Martínez, L.J.; Dugoujon, J.-M.; Novelletto, A.; Cuesta, P.; Pereira, L.; Calderón, R. Human genomic diversity where the mediterranean joins the atlantic. *Mol. Biol. Evol.* **2020**, *37*, 1041–1055. [CrossRef] [PubMed]

29. This, P.; Lacombe, T.; Thomas, M.R. Historical origins and genetic diversity of wine grapes. *Trends Genet.* **2006**, *22*, 511–519. [CrossRef] [PubMed]

30. Arroyo-García, R.; Ruiz-García, L.; Bolling, L.; Ocete, R.; López, M.A.; Arnold, C.; Ergul, A.; Söylemezo˝lu, G.; Uzun, H.I.; Cabello, F.; et al. Multiple origins of cultivated grapevine (*Vitis vinifera* L. ssp. sativa) based on chloroplast DNA polymorphisms. *Mol. Ecol.* **2006**, *15*, 3483–3861. [CrossRef]

31. Ramos-Madrigal, J.; Runge, A.K.W.; Bouby, L.; Lacombe, T.; Samaniego Castruita, J.A.; Adam-Blondon, A.-F.; Figueiral, I.; Hallavant, C.; Martínez-Zapater, J.M.; Schaal, C.; et al. Palaeogenomic insights into the origins of French grapevine diversity. *Nat. Plants* **2019**, *5*, 595–603. [CrossRef] [PubMed]

32. Plass, M.; Solana, J.; Alexander Wolf, F.; Ayoub, S.; Misios, A.; Glažar, P.; Obermayer, B.; Theis, F.J.; Kocks, C.; Rajewsky, N. Cell type atlas and lineage tree of a whole complex animal by single-cell transcriptomics. *Science* **2018**, *360*, 6391. [CrossRef] [PubMed]

33. Loeffler-Wirth, H.; Binder, H.; Willscher, E.; Gerber, T.; Kunz, M. Pseudotime dynamics in melanoma single-cell transcriptomes reveals different mechanisms of tumor progression. *Biology* **2018**, *7*, 23. [CrossRef]

Article

Developmental scRNAseq Trajectories in Gene- and Cell-State Space—The Flatworm Example

Maria Schmidt *, Henry Loeffler-Wirth and Hans Binder

IZBI, Interdisciplinary Centre for Bioinformatics, Universität Leipzig, Härtelstr. 16–18, 04107 Leipzig, Germany; wirth@izbi.uni-leipzig.de (H.L.-W.); binder@izbi.uni-leipzig.de (H.B.)
* Correspondence: schmidt@izbi.uni-leipzig.de

Received: 3 September 2020; Accepted: 14 October 2020; Published: 16 October 2020

Abstract: Single-cell RNA sequencing has become a standard technique to characterize tissue development. Hereby, cross-sectional snapshots of the diversity of cell transcriptomes were transformed into (pseudo-) longitudinal trajectories of cell differentiation using computational methods, which are based on similarity measures distinguishing cell phenotypes. Cell development is driven by alterations of transcriptional programs e.g., by differentiation from stem cells into various tissues or by adapting to micro-environmental requirements. We here complement developmental trajectories in cell-state space by trajectories in gene-state space to more clearly address this latter aspect. Such trajectories can be generated using self-organizing maps machine learning. The method transforms multidimensional gene expression patterns into two dimensional data landscapes, which resemble the metaphoric Waddington epigenetic landscape. Trajectories in this landscape visualize transcriptional programs passed by cells along their developmental paths from stem cells to differentiated tissues. In addition, we generated developmental "vector fields" using RNA-velocities to forecast changes of RNA abundance in the expression landscapes. We applied the method to tissue development of planarian as an illustrative example. Gene-state space trajectories complement our data portrayal approach by (pseudo-)temporal information about changing transcriptional programs of the cells. Future applications can be seen in the fields of tissue and cell differentiation, ageing and tumor progression and also, using other data types such as genome, methylome, and also clinical and epidemiological phenotype data.

Keywords: pseudotime trajectories; transcriptomic landscapes; differentiation of tissues; planarian; machine learning; self-organizing maps; single cell RNA sequencing

---

1. Introduction

Genome-wide single cell transcriptomics experiments provide snapshot data, which resolves the molecular heterogeneity of cell cultures and tissues with single cell resolution under static conditions [1,2]. These measurements are cross sectional and lack explicit time-dependent, longitudinal information about the developmental dynamics of each individual cell. Given that each cell can be measured only once, one needs models and computational methods to deduce developmental trajectories on cellular level and changes in underlying molecular programs from these static snapshot data. Such methods were developed in order to quantify transcriptional dynamics such as cell differentiation or cancer progression by using the concept of pseudotime (pt) [3–6]. The pt model assumes that single cell transcriptomes of different cells can be understood as a series of microscopic states of cellular development that exist in parallel at the same (real) time in the cell culture or tissue under study. Moreover, the model assumes that temporal development smoothly and continuously changes transcriptional states in small and densely distributed steps so that similarity of transcriptional characteristics can serve as a proxy of time. Here the pt represents the similarity measure used.

It scales development using values between zero and unity for the start and end points, respectively. Pt methods typically project the high-dimensional molecular data on to a space of reduced dimensions by (non-)linear transformations. In reduced dimensional space the cells were then aligned along a trajectory scaled in units of pt where a large variety of projection algorithms can be applied (see, e.g., [7–9]). A recent benchmarking study identified more than 70 pt-trajectory interference methods. About 45 of them were explicitly evaluated using criteria such as cellular ordering, topology, scalability, and usability [10]. Each method has its own characteristics in terms of the underlying algorithm, produced outputs, and regarding the topology of the pt trajectory. Methods make either use of pre-defined, fixed path topologies such as linear [3,11], cyclic, or branched [4,12,13] or they infer the topology from the data, e.g., as connected or disconnected graphs [12,14,15]. Most methods aim at inferring continuous cell state manifolds. To achieve this they transform single-cell data to graphs representing the individual cells as nodes, which are then connected by edges that reflect pairwise gene expression similarities. Such graph-based analyses are useful because they convert a set of isolated measurements of single-cell transcriptomes into a connected structure, which can then be analyzed using a rich set of mathematical methods for construction and visualization of the state space manifold and for (pseudo-)temporal analysis (see [16] and references cited therein). Method's performance depends on the trajectory type, dimensions of the data, and prior information where however often little is known about the expected trajectory. Notably, also different kinds of network studies aimed at inferring trajectories as directed graphs, e.g., in the context of metabolic flux analyses ([17] and references cited therein).

Hence, pt trajectories refer to ordered series of cell states. Alterations of activities of selected genes or gene sets along these trajectories then provide pt profiles of gene expression, which represent x-y plots depicting the expression levels as a function of pt [18]. They characterize (pseudo-)temporal changes of cellular programs upon development and can proceed, e.g., in a switch-like or in a more continuous fashion, or they can upregulate in intermediate, transient states [19]. Accordingly, molecular developmental characteristics can be split into two orthogonal views, namely focusing either onto the cells as the functional unit or onto molecular programs as changes of function independent of the associated cell state(s). Both aspects are closely related but not identical because development into different cell types can be driven by the same or by different molecular processes and, vice versa, different programs can associate with one or multiple cell types. For example, co-evolution of tumor cells and their microenvironment involves different cell types and states, which are expected to show co-regulation in gene-state space and potentially could support inference of the development of molecular communication networks between different cell types (see [20,21] and references cited therein). Trajectories in cell-state space are addressed by the numerous pt-approaches mentioned above while elaborated methods and applications for visualization and analytics of developmental paths in gene state do not exist to the best of our knowledge.

We previously developed the expression portrayal method that visualizes multidimensional, whole genome expression landscapes in terms of two-dimensional images making use of self-organizing map (SOM) machine learning [22,23]. Such landscapes are typically characterized by clusters of genes, so-called spot modules, concertedly overexpressed under certain condition such as cell types or developmental stages. The self-organization properties of the method arranges the spot-modules of similar expression patterns closer to each other than more dissimilar ones [22]. Hence, SOM machine learning arranges spot-clusters of co-regulated genes according to mutual similarities of their expression profiles in an ordered fashion resembling paths in gene space. In other words, SOM accomplishes the task of sorting features in gene-state space in analogy to the task of sorting cells according to mutual similarities of their transcriptomes along pseudotime trajectories in cell-state space.

We have previously shown that the arrangement of spots indeed reflects dynamics of different processes. For example, melanoma progression from naevi to metastatic tumors has been analyzed by combining SOM portrayal and pt-scaling of bulk melanoma transcriptomes [24] and single-cell data of melanoma cell cultures [19,25]. Another study addressed organoid development of liver buds using a similar approach [26]. The ability to deduce dynamic information in SOM gene-state space from cross-sectional data was supported by (real) time-resolved SOM-analyses of the longitudinal transcriptomic changes along the yeast cycle [27], of dynamic changes of urine proteomics during space flight simulations of human individuals [28] and, earlier, equilibration of cellular states after external perturbations into stable attractor states [29,30]. In these applications SOM provided trajectories in the molecular landscapes. The studies so far associate with relatively simple linear, circular, bifurcation-like, and converging processes. Furthermore, the trajectories in feature-space were deduced more on a qualitative level without underpinning them with appropriate computational methods. Finally and most importantly, single cell "omic" technologies strongly evolve and now became standard in many applications beyond the flatworm example used here. Presently, many ten-thousands of cells are sequenced in parallel, quality of transcriptomic data is improved, technology is refined, e.g., in direction of spatially resolved transcriptomics, and extended by single-cell proteomics, genetics, and chromatin accessibility methods.

In this publication, we aim at further developing the idea of constructing developmental trajectories in SOM-feature state space from cross-sectional data in order to supplement pt trajectories in cell-state space with such trajectories describing developing molecular programs. Our efforts should be understood as a first step in this direction as a proof of principle approach. It extends previous work by explicitly extracting "similarity paths" in the expression landscape provided by the SOM method. In other words, we supplement our portrayal of static molecular landscapes by (pseudo-)dynamical views in analogy to pt approaches in cell-state space.

Making use of topological features of the expression landscape we also process so-called RNA-velocities as an independent approach. RNA-velocity analysis directly "forecasts" the transcriptional state of a cell based on the relation between spliced and unspliced mRNA [31,32]. We here apply this concept to gene-state space to derive local gradients of transcriptional activity pointing toward attractors of gene activity along RNA-velocity paths. Hence, as the second novelty we project RNA-velocity information into expression portraits.

We selected "flatworm" single cell RNAseq data as an illustrative example. Planarian transcriptomics was well studied as a first application of scRNAomics to developmental dynamics in a whole complex animal [29]. This data enabled to reconstruct multibranched lineage relationships of cell differentiation from stem cells into different tissue types and to identify gene sets, which program the lineage tree. The animal overall consists of several tissue types, so that multibranching of developmental trajectories is sufficiently diverse.

Our publication is organized as follows (as shown in Figure 1): first, we introduce both cell- and gene-state centric views on development of Planarian tissues. Second, we characterize the SOM-expression landscape and extract differentiation paths into selected tissue types. For comparison with cell-state space we made use of the URD-multibranched diffusion pt method. Finally, we mapped RNA-velocities into SOM to derive vector fields of changes of RNA abundance. Overall, the SOM-based gene expression analysis of developmental trajectories complements cell-trajectories by more clearly disentangling involved molecular programs.

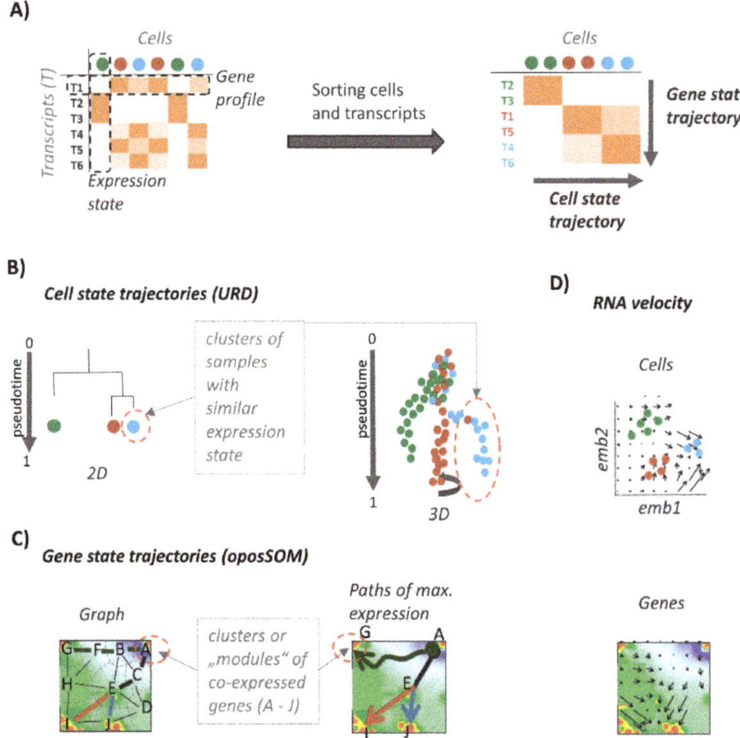

Figure 1. Analysis of developmental paths (schematic overview). (**A**) Single-cell transcriptome data are provided as a matrix of cells (columns) versus transcripts (rows). Similarity sorting along genes and cells is applied to obtain gene- and cell-state trajectories, respectively. (**B**) Multibranched cell-state trajectories order cells along a similarity measure called pseudotime (pt). They are visualized either as hierarchical tree or as three-dimensional manifold using, e.g., "URD"-plots [29]. They reflect transcriptional changes upon development and differentiation of cells. (**C**) The gene-state trajectories were generated in self-organizing map (SOM)-space making use of a network formed between clusters of co-regulated genes marked with letters in the figure. Trajectories follow paths of maximum expression linking source (stem cells) and sink (differentiated tissues) nodes along the edges of the net. (**D**) RNA velocity analysis delivers vector fields in cell and gene phase space. Each vector points in direction of increasing transcript abundance either between cell states or between gene states. In fact, RNA velocity forecasts the "future" transcriptional state of each gene. The RNA-velocity vector of the cells is formed by the RNA-velocities of all genes as components while the RNA-velocity vector in gene space is composed by the RNA-velocities of all genes in the respective local gene cluster.

2. Materials and Methods

2.1. scRNA Seq Data of Planarian Development

Single-cell RNA Seq data of a whole planarian Schmidtea mediterranea were taken from the publication of Plass et al. [29] as read-counts (in units of RPKM) of more than 28,000 transcripts per cell. Originally these data were measured by highly parallel droplet-based single-cell transcriptomics, Drop-seq [33], to study different cell types and progenitor stages present in adult planarians. In total 21,612 cells were captured in 11 independent experiments. The data set corresponds to five wild-type samples (10,866 cells), two RNA interference (RNAi) samples (3314 cells), a high-DNA content G_2/M population corresponding to cycling planarian stem cells (typically defined as x-ray–sensitive

"X1 cells"; 981 cells), and three wild-type regeneration samples (6451 cells) [34,35]. As in the original publication [29], we analyzed all single-cell data sets together. Cell types were assigned to the previously annotated 51 cell clusters. Cell type identities of neoblast, neural, epidermal, secretory, muscle, gut, and protonephridia cells were further elucidated by examining known marker genes based on previous knowledge [29]. We applied the same color scheme as in the figures shown in [29] throughout this publication for direct comparison.

2.2. Single Cell SOM Portrayal

For transcript-related analysis we applied the "Single-Cell R Analysis Toolkit" available as R-package "scrat" [22,23,27]. In brief, it works as follows: Transcript-level expression data were log-transformed and centralized into log-fold changes with respect to the ensemble average of each transcript, log FC (cell) = log expression (cell)—mean_log expression (all cells). Then, SOM neuronal network machine learning translates the high-dimensional expression data of N = 28,065 gene transcripts into K = 3600 metagene expression data per cell. Implementation of SOM and optimal parameter settings were addressed previously [27]. Each metagene represents a "micro"-cluster of co-expressed genes showing mutually similar expression profiles across the samples. Metagenes were arranged in a 60 × 60 two-dimensional grid coordinate system and colored according to their expression level in each sample thus generating an image of the transcriptome of each single cell. The resolution of 60 × 60 metagenes was chosen to optimally resolve expression modules as shown previously. Default coloring of metagenes (red to blue for maximum to minimum expression, respectively) scales with log-expression values [22,36]. Mean portraits of the transcriptome of a cell type were calculated by averaging metagene expression values over all single cell portraits of the respective type.

Metagenes of mutually correlated profiles cluster together forming "spot-like" red and blue areas of over- and under-expression in the portraits because of the self-organizing properties of the SOM. These "spot"-modules were detected using distance-metrics (D-map) clustering [37]. They extract groups of co-expressed genes showing high expression in specific cell types. However, the spots cover only part of the SOM and thus only part of the genes. For "space-filling" segmentation of the SOM we applied k-means clustering to all metagenes. This k-means clustering stratified the 3600 metagenes into 32 clusters where 11 of them can be assigned to the spots, 21 collect virtually low variant transcripts, and one cluster collects almost invariant genes (see below). The choice of the number of metagenes, clustering criteria, and methods for segmenting SOM were extensively studied and described previously [22,27]. In short, a number of several thousand metagenes are large enough to resolve transcriptional landscapes in an unsupervised fashion in order to identify dozens of different states with sufficient resolution. Larger numbers of metagenes have virtually no effect on the results. The number of spots represents an intrinsic property of the data. It is obtained in an unsupervised way by searching for modules of co-expressed genes by applying different algorithms based on overexpression or D-map criteria. Typically their number is roughly comparable with the number of transcriptomic states as has been shown in previous applications [25,36,38]. It is expected to vary between ten and twenty in the flatworm example. For low-variant modules a number exceeding the number of spot modules by a factor of two is appropriate to segment the map with sufficient resolution. A number too small (e.g., five or less) will provide a too coarse resolution of the paths while higher numbers (e.g., more than thirty) do not change the results and, moreover, typically produce "empty" clusters without functional information.

The spot-patterns obtained represents a characteristic fingerprint of each particular sample. This "logFC" scale is used as default color scale of the portraits (see also [22]). It enables to identify well-separated areas of overexpression (red color) as described above. In addition, we applied a so-called "loglogFC" scale for alternative coloring. It applies the signed logarithm of the absolute value of logFC, loglogFC = sign log(abs(logFC)), where "sign" equals the sign of logFC. First of all, the "loglogFC" scale identifies a "coast-line" which separates areas of overexpression (FC > 1) in red from areas of underexpression (FC < 1) in blue. LoglogFC portraits overall better resolve subtle changes of the expression landscape in low expression areas.

2.3. Cell Clustering, Cell State Trajectories, and Pseudotime

For an overview, we visualized similarities between transcriptomes of planarian single cells using t-distributed stochastic neighbor embedding (tSNE) algorithm [39]. Pseudotime (pt) analysis was performed by means of simulated diffusion paths through the cells in similarity space as implemented in URD [40]. It distributes the scRNAseq transcriptomes along a branched tree-structure reflecting different trajectories of cell development. We assigned the root of differentiation to planarian stem cells. Then URD calculates the trees by constructing a k-nearest-neighbor graph and assigns a pseudotime (pt) by simulating a diffusion-like process from the root to the tip of the graph. The pt scales similarities between the cell transcriptomes using a value between pt = 0 at the start to pt = 1 at the tip of largest distance. Overall application of URD delivers a branched trajectory from one root to several tips.

2.4. Gene State Trajectories in SOM Space

Gene-state trajectories were calculated by applying the maximum flow algorithm [41] to transcriptomic data in metagene space as provided by SOM. To generate these trajectories, we first segmented metagene space into space-filling k-means clusters as described above. Then they were transformed into a network with edges linking all neighboring clusters. The weights (called capacities) of the edges were given by the cumulative mean loglog FC value of the two adjacent clusters linked by the edge. Next, we defined a source cluster (starting point) and the sink cluster (end point) of the differentiation process, e.g., between genes upregulated in undifferentiated stem cells and genes upregulated in differentiated tissues. The maximum flow-algorithm connects source and sink along the edges providing the maximum cumulative weights. This path then defines the gene-state trajectory ensuring maximum cumulative expression between source and sink. Alternatively we made use of different portraits between differentiated tissue and stem cells to generate gradient trajectories along topographic least cost paths from source to sink spots using the R-program topoDistance [42].

2.5. RNA-Velocity Dynamics

The so-called RNA-velocity is the time derivative of gene expression. It is positive for increasing and negative for decaying transcript abundance. It can be evaluated from the relative amounts of unspliced and spliced mRNAs of each gene taken from RNAseq data, their ratio under steady state conditions and assuming a constant splicing rate of all genes [31]. The mRNA-velocities of all genes form a multidimensional vector pointing in direction of the overall increment of mRNA of the cell and thus predicting its future expression state. In metagene space as provided by SOM the mRNA velocity of a metagene is given by the multidimensional vector with the mRNA-velocities of all single genes included in the metagene cluster as components. It points toward the future state of the metagene in the expression landscape, or, in other words, it forecasts the expression change of the metagene. Overall, the metagene-related mRNA-velocities thus provide vector fields pointing toward local attractors of maximum transcription (see below). The vector field obtained is then transformed into trajectories using RNA-velocity [31]. It summarizes the metagene-related RNA-vectors into "flow"-vectors toward the sinks (or attractors) of transcription (see below).

2.6. Function Analysis

We applied gene set analysis to the lists of genes located in each of the modules to discover their functional context using the planMine database [43]. planMine takes the Gene Ontology (GO) information for a transcript from the NCBI database for RefSeq proteins that show strong Blast homology with Planarian transcripts [43]. Gene set maps complement this analysis by visualizing the position of the genes of a set within the SOM grid. According to their degree of accumulation in or near the spot modules, one can deduce their potential functional context [22].

2.7. Methods Availability

Trajectory generation and visualization is implemented as R-package oposSOM.PT (https://github.com/mschmidt000/oposSOM.PT). It should be used together with the R-package oposSOM [19,20] (https://github.com/hloefflerwirth/oposSOM) in order to generate trajectories upon running the program.

3. Results

3.1. Portrayal of Developing Single Cell Transcriptomes

For the comprehensive cartography of cell states upon tissue development of planarian, we generated SOM transcriptomic portraits of all nearly 22,000 cells studied. Then, we summarized them into mean portraits of each of the 51 cell types defined previously [29] by averaging all single cell-portraits of each type. tSNE plots were used for an overview of transcriptional diversity of the SOM portraits (Figure 2A). Cells were grouped into 51 clusters in accordance with the classification and coloring scheme used previously [29] (see legend in Figure 2E). Our SOM-based results are in line with that of Plass et al. [29] showing one large cluster of neoblast stem-cells (grey color) which is surrounded by clusters of differentiated cells of different tissues (dark color tones) and clusters of progenitor cells (brighter colors) in between (Figure 2A). The "differentiation" axis points roughly from the stem cells in the right part toward differentiated cells in the left. The SOM expression portraits of the cell types are characterized by red and blue spots at specific positions. They refer to clusters of concertedly over- and under-expressed genes, respectively. For example, neoblasts show a red overexpression spot in the upper right corner of the map, while differentiated muscle and goblet cells show overexpression spots in the left and middle bottom area, respectively (Figure 2B). The respective progenitor cell express a virtually bimodal spot patterns consisting of a mixture of the spots observed in the undifferentiated and differentiated cells, respectively. One can understand differentiation as a series of activated cellular programs, which become evident as path in the expression landscape linking the stem cell spot with the spot observed in differentiated cells. Examples of such developmental trajectories of changing gene states were shown in Figure 2D for muscle, gut, epidermal, and neuronal cells obtained by means of the maximum flow algorithm (see Materials and Methods section). Changing cell states upon differentiation were visualized using URD-clustering. This method sorts cells according to mutual similarities along pseudotime trajectories, which are specific for different developmental paths (Figure 2C).

In summary, we here introduced two complementary views on developmental trajectories, namely as paths describing the sequence of cellular states passed between stem cells and differentiated tissue cells on one hand and as the sequence of major gene regulatory programs activated upon differentiation on the other hand. SOM-space visualizes the latter aspect. It enables studying functional aspects of development in gene space.

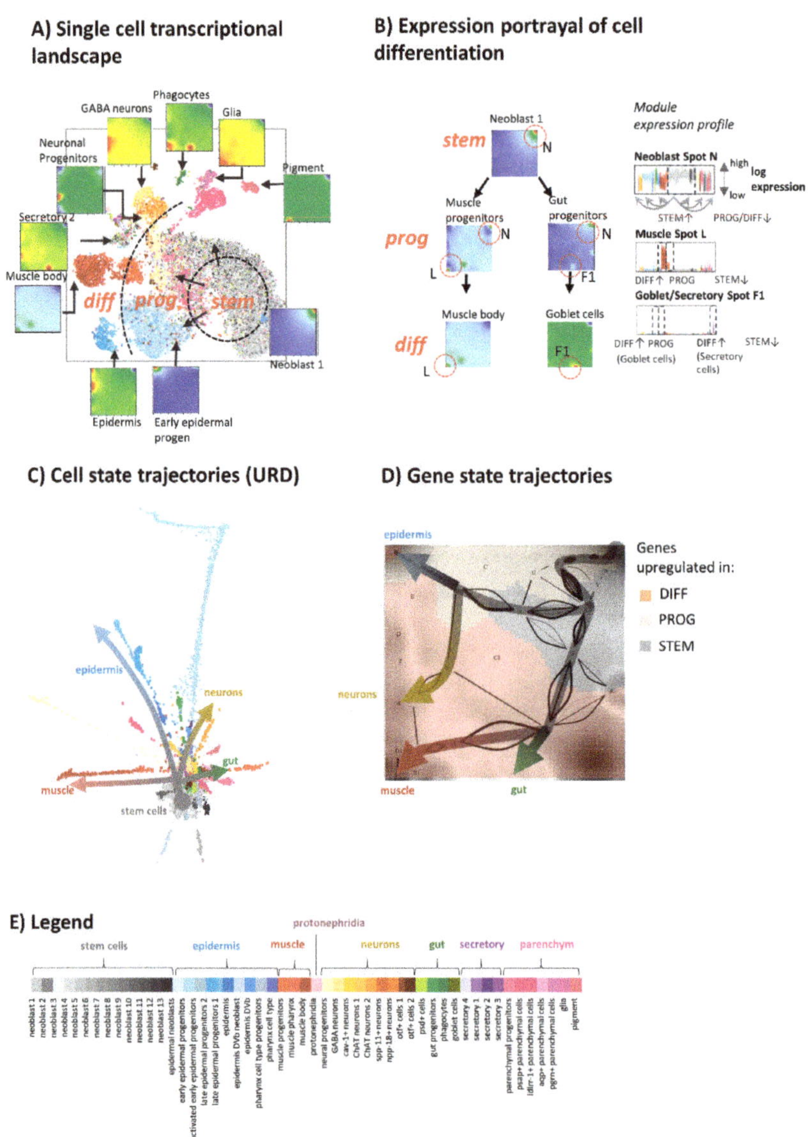

Figure 2. SOM portrayal of the developing planarian single cell transcriptome: (**A**) Single-cell resolved tSNE plot (cell types are color-coded as shown in the legend in part E). SOM expression portraits of selected cell types provide a glimpse on cell and gene transcriptional space. (**B**) Expression portraits of stem cells and of differentiating progenitor and differentiated muscle body and a goblet cells show overexpression spots at different positions as indicated by the dashed circles. The spot profile reveals specific up regulation of the spots in the respective tissue specific cells. (**C**) The URD [40] plot resolves multibranched single-cell developmental manifolds. Paths of selected tissue types are indicated by arrows. (**D**) Gene activation trajectories of cell differentiation from stem cells into four different tissues were calculated by means of maximum flow on SOM embedding. (**E**) Legend of colors used for cell-type assignment (see also [29]). The color code is used throughout the paper.

3.2. The Transcriptome Landscape of Planarian

Next, we further characterized the topology of SOM expression landscape and its functional context in detail. In total, we segmented the 3600 metagenes used for the SOM training into 32 expression modules by applying k-means clustering to the metagene's expression profiles. The modules were labeled by capital letters "A–F1" (Figure 3A). A summary of the top expressed marker genes in each of the modules, enriched functional gene sets, and samples showing upregulation are provided in Table 1. The modules are divided into 11 highly variant expression clusters (A, D, F, J, L, N, R, B1, D1, E1, F1) and into 21 less variant clusters. The former ones form "spot-like" red areas along the periphery of the SOM which specifically upregulate in the different tissues (Figure 3C). The low variance clusters were found in the middle and right lower part of the map. About 65% of all genes are virtually invariant and accumulate in cluster A1 evident as blue area in the summary and variance maps (Figure 3B). Hence, the topology of the SOM splits into one developmental source spot-collecting genes overexpressed in stem cells, into a series of developmental sink spots-collecting genes upregulated in differentiated tissues and into an intermediate transition region in between. Further, SOM identified an area collecting virtually invariant genes and spot R of unknown origin. Presumably, it can be attributed to methodical effects systematically disturbing parts of the expression patterns.

The landscape is governed by the amplitudes and particular shapes of the gene expression profiles. They are shown around the map for selected modules together with gene maps indicating the positions of gene referring to specific Gene Ontology (GO) terms (Figure 3A). For example, genes of the GO-term "Muscle structure development" accumulate in and near the "muscle" module L upregulated in muscle tissue this way confirming its functional impact. The heatmap in Figure 3C visualizes the profiles of the less variant clusters. Overall, these profiles split into three major strata either upregulated in stem cells (marked with grey color), progenitor cells (light apricot), or differentiated cells (apricot), respectively (Figure 3A–C). The progenitor cell clusters can be further divided into "earlier progenitors" activated still in stem cells and into "later progenitor cells" not showing activation in stem cells. Marker genes of differentiation between the different planarian cell types (Table S1 and [44]) map into or near the modules upregulated in the respective cell types (Figure 3D). For example, the epidermal module J includes the epidermal gene markers prog-1 or agat-1 while the stemness module N contains the stem cell marker Smedwi-1 and tub-α1. Overexpression analysis of selected reference sets of co-regulated genes also well agreed with our modules. They further specify functional interpretation, e.g., in terms of activated cell cycle phases within the stem cell compartment (see Figure S1B–E).

Note that spots and underlying cellular programs are activated in different cell types (see the profiles in Figure 3A). For example, cell-cycle-related transcription is virtually active in all cell types however to a different extent (Figure S1D). These genes locate mostly in spots W and V near the stemness module N. Other common programs are more selective: genes related to morphogenesis accumulate in spot F upregulated in epidermal and muscle tissues. Such genes activated in several differentiated tissues overall accumulate in a stripe-like area near the lower left corner of the map including spots F, Y, and M (Figure 3A,D).

Gene set enrichment analysis using the database planMine for mining flatworm genomes and transcriptomes [43] provided detailed functional information associated with each cell type (Table 1). For example, the top enriched sets in the "muscle module" L are the GO-terms collagen fibril organization, sarcomere organization, and muscle tissue development while the "stem cell module" N enriches terms such as rRNA processing, ribosomal subunit biogenesis, and RNA processing, reflecting the high biogenetic and metabolic activity of stem cells. In summary, the diversity of the scRNAseq data decomposes into a set of one dozen clusters of co-regulated genes of defined functional context. Along the differentiation axes the cells split into neoblasts, early, and late progenitors and differentiated tissue cells according to their transcriptomic patterns.

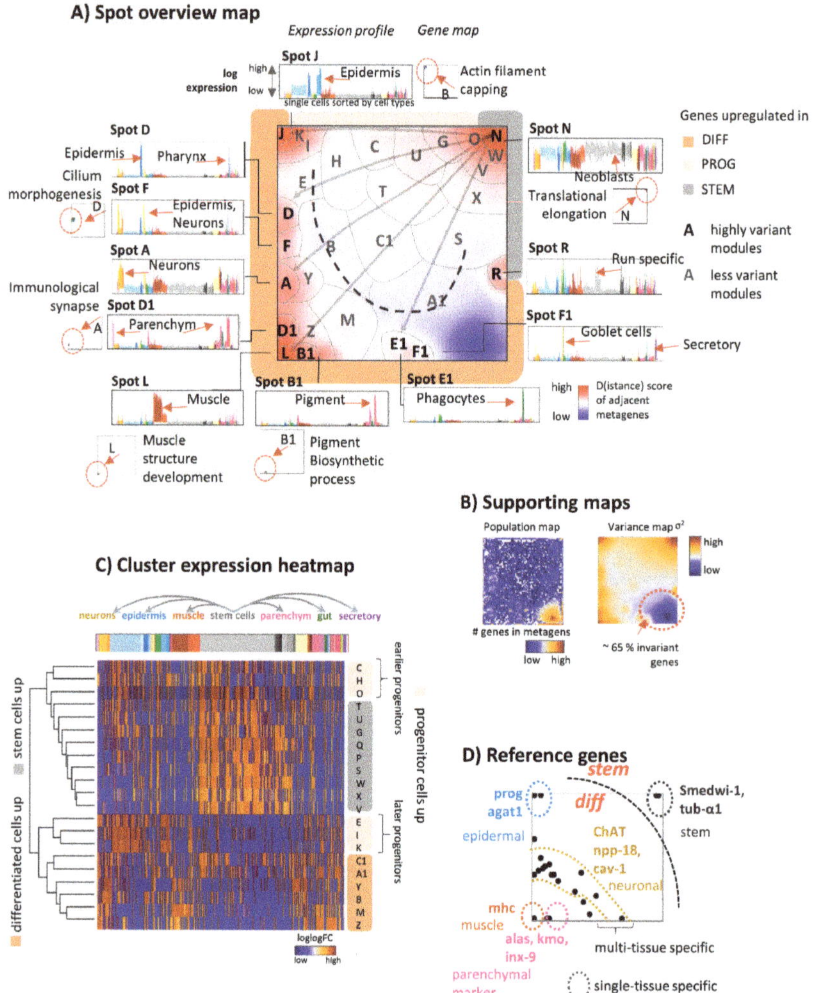

Figure 3. The single-cell transcriptomic landscape of the planarian Schmidtea mediterranea: (**A**) The module overexpression map provides an overview about modules of co-expressed and highly variant genes. They mainly arrange along the edges of the SOM and are labeled with bold capital letters A, D, F, J, L, N, R, B1, D1, E1, F1. The grey arrows illustrate selected developmental paths by linking the stemness spot N with the respective "tissue" spots. Expression profiles (gene set enrichment Z-score, GSZ-scale) and gene maps of selected functional gene sets underpin the functional context of these modules. They reveal specific up- and downregulation of the gene sets in different tissues (red arrows). The gene maps indicate the location of the genes by dots. Accumulation of genes in and near module areas are indicated by red arrows and dashed ellipses. (**B**) Population and variance maps visualize the number of genes in each metagene (log-scale) and the variance of metagene expression, respectively. About 65% of all genes studied show virtually invariant expression. They accumulate in the blue area in the right lower corner as indicated. (**C**) Expression profiles of the modules of low variant expression (labelled with non-bold letters in part A) are shown as heatmap. The arrows above the heatmap point in the direction of differentiation from stem cells into different tissues. The spot modules group roughly into three clusters as indicated in the right part of the heatmap and along the edges of the SOM in part A. (**D**) Location of key reference genes are shown by dots (see also Figure S2 and Table S1). They are mostly found in or near the spots which upregulate in the respective tissues.

Table 1. Overview of module-wise enriched gene sets and marker genes of planarian sc RNASeq data.

Cluster	Genes	Name	Enriched BP Gene Sets (log p Enrichment)	Marker Genes	Tissue
A	235	Neurons 1	Neurotransmitter secretion (-6) Synaptic vesicle localization (-5) Neurotransmitter transport (-5) Regulation of neurotransmitter levels (-5) Synaptic vesicle transport (-4)	pc2, ChAT, gpas	cav-1+ neurons, ChAT neurons 1, ChAT neurons 2
B	1314		Phosphorus metabolic process (-4) Phosphate-containing compound metabolic process (-4) Homopholic cell adhesion via plasma membrane adhesion molecules (-4) Cell-cell adhesion via plasma-membrane adhesion molecules (-4) Cell communication (-3)	GABRB3 (dd21541), CAVII-like, th	
C	206				
D	265	Pharynx		vim-1,VIT (dd1071), NPEPL1 (dd181)	Epidermis, Pharynx
E	383			1.G9.2, ifn	
F	370	Neurons 2	Microtubule-based movement (-18) Microtubule-based process (-15) cilium morphogenesis/organization/assembly (≥ -14) movement of cell/subcellular component (-9)	pkd2l-2, rootletin (dd6573), cav-1	cav-1+ neurons, GABA neurons
G	218				
H	330			Post-2c	
I	228			nb.22.le	
J	111	Epidermis	actin filament capping (-2) negative regulation of actin filament depolymerization (-2)	prog-2, prog-1, agat-3	Epidermis, late epidermal progenitors 1/2
K	100			pds	
L	150	Muscle	muscle structure development (-9) tissue development (-8) actin filament-based process (-7) anatomical structure development (-6) muscle organ development (-6)	COL4A6A (dd2337), collagen, COL21A1 (dd9565)	Muscle body
M	1032			cali, if-1, HSPG2 (dd8356)	
N	121	Stem cells	translational elongation and termination (≥ -79) cotranslational protein targeting to membrane (≥ -73) protein targeting to ER (-72) Peptide and amid biosynthetic process (≥ -71)	smedwi-1, dd_6998	Stem cells
O	143				
P	114			bruli	
Q	44			vasa-1	
R	119	Run specific	Unknown origin, presumably due to batch effects		Run specific cells
S	974			dd_5560, SAMD15 (dd19710), wntP-3	
T	404			sp-5	
U	235				
V	265			TYMS, gH4	
W	146		Gene expression (-8) Nucleic acid metabolic process (-7) Cellular nitrogen compound metabolic (-6) Cellular macromolecule metabolic process (-5) Nucleosome organization (-5)		
X	398			dd_13666	
Y	445			GLIPR1 (dd210), npp-18	
Z	575			TMPRSS9 (dd7966), CTSL2 (dd582), gata 4/5/6	
A1	18239		G-protein coupled receptor signaling pathway (≥ -47) cell communication (-12) Signaling (-11) single organism signaling (-11)	glipr-1, PI16 (dd940), ASCL4 (dd1854)	
B1	181	Pigment	organonitrogen compound catabolic process (-6) tyrosine catabolic process (-5) small molecule metabolic process (-4) response to transition metal nanoparticle (-4) L-phenylalanine catabolic process (-4) erythrose 4$-$	pgbd-1, PSAPL1 (dd1706), KMO (dd7884)	Pigment cells
C1	458				
D1	262	Parenchym	regulation of intracellular signal transduction (-3) negative regulation of antigen receptor-mediated signaling pathway (-3)	ctsl2, CTP, PCN, dd_Smed_v6_3_0, PSAP	aqp+ parenchymal cells, pgrn+ parenchymal cells, psap+ parenchymal cells
E1	226	Phagocytes		mat	Phagocytes
F1	725	Goblet, secretory cells			Goblet and secretory cells

3.3. Portrayal of Developmental Paths

In the next step, we aimed at characterizing gene-state expression dynamics during differentiation. For a simple first view, we ordered the expression portraits of the cells of epidermal differentiation along the linear, unbranched pseudotime path taken from Plass and colleagues [29] (Figure 4A). Hereby the stemness module is highly expressed in the portraits of stem cells, of epidermal neoblast, as well as in epidermal progenitors. Its amplitude however progressively decays in the portraits of early and late epidermal progenitors but then markedly decays in the portraits of differentiated epidermal cells. On the other hand, the expression of the epidermal module progressively increases in an antagonistic fashion. Notably, this switching affects about 1000 genes included in these two modules. We replot the default "logFC"-scaled portrait into double-logarithmic "loglogFC" scale in order to better visualize subtle expression changes upon differentiation. The red overexpression region "flows" along the upper border of the map in direction from the stemness "source" spot at the right toward the epidermis "sink" module at the left. A gene-state trajectory of epidermal differentiation was computed using the maximum flow algorithm [10] as described in the methods section. Briefly, the trajectory is obtained as directed graph linking the source and sink module via the path of maximum cumulative expression between the adjacent modules along the path (Figure 4B). As expected, the trajectory follows the right-to-left shift of the red overexpression region seen in the portraits upon differentiation. The expression of stemness genes decays along the trajectory and it increases for genes involved in epidermis differentiation (Figure 4C). Gene-state trajectories of differentiation, pt profiles, and associated expression portraits of parenchymal, neuronal, muscular, and gut tissues are shown in Figure 4D. Hereby, differentiation is described by trajectories starting in the common stemness mode and ending in the respective tissue-specific expression modules.

As an alternative method we applied a lowest cost path algorithm to the difference map between stemness and differentiated tissues (Figure 4E) [42]. The trajectories obtained closely resemble those obtained by means of the maximum flow algorithm. Interestingly, the gene states move "downhill" along the differentiation trajectories in the difference SOM, a picture, which resembles the epigenetic "Waddington" landscape [18,45]. This landscape was introduced as a conceptual model to illustrate differentiation of tissues from stem cells. Note that the SOM-landscape is a "real" expression data landscape, which this way supports the abstract Waddington concept of developing cells in gene-state space. In summary, our approach identifies cascades of gene modules changing expression upon differentiation in a tissue-specific (pseudo-)temporal order. In SOM metagene space it appears as "lava lamp"-like flows of upregulated genes, which can be summarized into gene expression trajectories directed from stemness toward differentiated tissue expression programs.

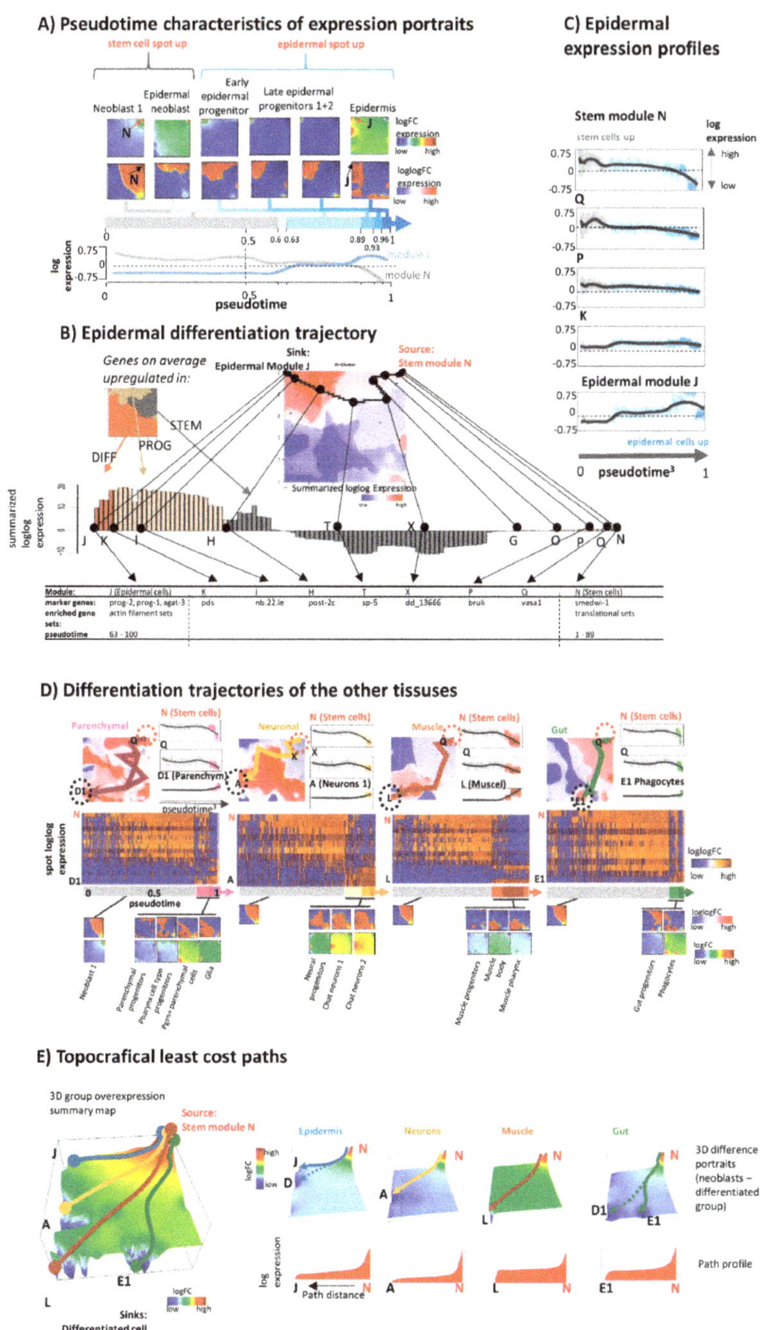

Figure 4. Gene expression trajectories of planarian tissue development. (**A**) Expression portraits (in log FC and loglog FC scales) of the epidermal lineage illustrate the change of module-patterns during

differentiation. The "loglogFC" scale better resolves subtle changes between over- and under-expression in red and blue, respectively. Alterations of expression patterns become evident as "lava lamp"-like flow of the red spot from the right to the left. The expression of stemness and epidermal spots N and J change in an antagonistic fashion along the pseudotime. (**B**) The gene-state trajectory of epidermal development links the stemness spot with the "epidermal" spots in the expression landscape shown as cumulative loglogFC SOM). The expression profile along the trajectory assigns genes and functions along the path. (**C**) Selected expression profiles of spots along the developmental path of epidermal cells switch from negative to positive slopes between spots P and K. (**D**) Developmental characteristics of four different planarian tissues. The row above shows SOM-trajectories and pt-profiles. Below, the heatmaps of expression modules characterize expression changes of the respective tissues upon development as a function of pt. The respective expression portraits are shown below. (**E**) Gene state developmental trajectories obtained from the difference of SOM between stemness and differentiated tissue states using "topoDistance" analysis show differentiation paths as in part A–D. Largest expression changes were observed in the "final" slope describing differentiation of tissues from progenitor states.

3.4. Development along Branched Trees

The unbranched pt approach applied in the previous subsection sorts the cells along one dimension independent of the cell type. A more realistic computational approach is provided by URD [40]. It splits the pt path of planarian tissue development into a tree-like, multibranched one. The algorithm applies diffusion-like random walks to search for graphs through cellular states. The obtained trajectory starts with a common trunk of neoblast cells for early pseudotime values at pt < 0.6 (blue color in Figure 5A,B). Subsequently it divides into tissue-specific branches at pt > 0.6 (yellow to red). Notably, the multibranched tree obtained by means of the URD algorithm reproduces the differentiated cell types identified by independent methods in the original publication [29]. Different visualizations of URD branching in terms of a 3D cell cloud (Figure 5A), of a path-tree or t-SNE plots (Figure 5B) enable inspection of paths under different perspectives. The tree describes differentiation into more than thirty final tissue types via 13 progenitor populations (Figure 5C). Initial stemness states stratify into a variety of neoblast and progenitor cells such as gut and early epidermal progenitors at pt = 0. Subsequently, the main branch splits the trajectory into sub-branches describing differentiation into tissues such as parenchymal, neuronal, proto-nephridial, and diverse epidermal and gut cells. Further branching separates late epidermal progenitors, pigment and epidermis cells, muscle body and muscle pharynx cells, and npp-18 and otf+ cell neurons.

The profiling of spot gene expression along the different sub-branches using URD-pseudotime (pt) as argument shows the decline of expression of the stem cell module "N" and the gain of the expression of the spot-modules referring to differentiated tissues as expected (Figure 5D). Differentiation proceeds rather step-wise which concerns variance of the data than continuously at different pt-values ranging from about pt = 0.1 (epidermal tissue) and 0.2 (muscle) to 0.5 (neurons) which suggests switching of the cellular machinery from stemness programs into programs of differentiated cells almost without intermediate expression states. Some tissues show indications of intermediate states (e.g., muscle, epidermis) in terms of more smooth transitions between undifferentiated source and differentiated sink states.

Next, we made use of previous stratification of planarian genes into 48 clusters of co-expressed genes, which were assigned to different tissues and tissue-combinations [29]. The tree in Figure 5E illustrates the obtained hierarchy and maps marker genes taken from [29] into the SOM. In general, one observes that genes upregulated in several tissues (upper part of the tree) locate more in the middle part of the SOM and of the intended trajectories while marker genes for single tissues are found in and near the respective tissue-related spots detected in our analysis. Overall, the spread of genes is relatively large, presumably because the number of 48 clusters used in [29] is insufficient to resolve the expression patterns underlying tissue differentiation. On the other hand, the location of part of the tissue markers slightly away from the sink nodes presumably reflects also the fact that they are activated in transition states as progenitors of more than one tissue type as visualized in the tree (Figure 5E). Tissue markers of our spot-analysis are listed in Figure 5E and Table 1. In summary, multi-branched pseudotime approaches such as URD provide an overview of tissue differentiation by sorting the cells into differentiation paths based on their mutual similarity relations.

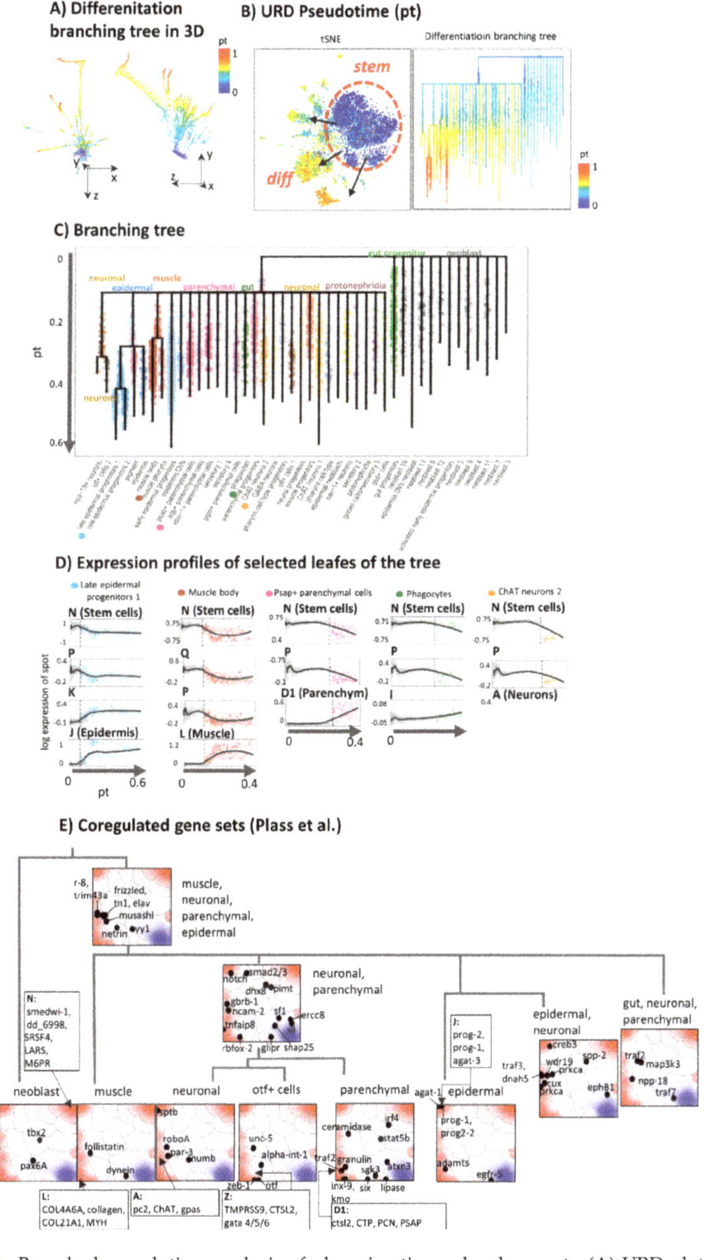

Figure 5. Branched pseudotime analysis of planarian tissue development. (**A**) URD plot [40] of branched differentiation in 3D is shown in two projections. The pseudotime is calculated as the average

number of diffusion steps required to reach each cell from the root cell (neoblast 1) [40]. (**B**) Pseudotime coloring of t-SNE plot and URD-branching tree. In pt-units epidermal tissues are most distant from the root. (**C**) Branching tree of cell differentiation. (**D**) Expression profiles of selected lineages along the branches of the URD differentiation tree. Samples are sorted along pseudotime (gray arrow). The black LOESS (locally weighted scatterplot regression)-curves serve as guide for the eye. Step-wise transitions between stemness and tissue transcriptional programs are shown by vertical dashed lines. (**E**) A tree of tissue-related gene clusters together with marker genes were taken from [29]. Accordingly, 48 co-regulated gene sets assign to Planarian tissues and combinations of them. Most of these genes locate not in the spots referring to fully differentiated tissues. The top five genes in these clusters according to our analysis are listed in the figure (see boxes and also Table 1).

3.5. RNA Velocity Trajectories

RNA-velocity analysis offers an independent approach of studying developmental dynamics. It calculates the change of mRNA abundance of each gene per unit of time in every single cell making use of the relation between spliced and unspliced mRNA and assumptions about the reaction kinetics of splicing [31]. RNA velocity is positive if the amount of mRNA increases. It refers to the situation when the amount of unspliced mRNA exceeds the amount of spliced mRNA compared with the steady state situation meaning that transcription and mRNA degradation are in equilibrium. Overall, one obtains a multidimensional RNA-velocity vector for each cell where each dimension refers to the rate change of transcription of one gene. The combination of velocities across genes then predicts the future expression state of the cell. In its original implication, the mRNA velocity vector points to direction of increasing mRNA abundance in sample space. For planarian single cell transcriptomics such plots show the expression state to which the cell is apparently moving in time [31].

Here we apply this concept also to expression space, where RNA velocity refers to the rate change of mRNA expression in each metagene-pixel of the SOM-space. It results in a vector whose components are given by rate changes of all individual genes contained in the metagene-cluster. The resulting RNA-velocity vector points in the direction of local mRNA rate changes in SOM space. Overall mRNA velocity analysis thus generates a metagene-based vector-field pointing toward increasing transcript abundance. It consequently points in the direction of "high" expression spot areas containing either genes upregulated in differentiated tissues such as epidermal (spot "J"), neuronal ("A", "Y" and "F"), and muscular ("L") tissues or genes upregulated in stem cells (module "N"). Hence, overexpression spots represent attractors of RNA-velocity (compare with Figure 3C). In turn, areas of invariant expression lack measurable RNA rate changes. Hence, RNA velocity analysis provides information about local slopes of mRNA abundance in gene expression space. Interestingly, RNA-velocity "vanishes" in the kernel-area of maximum expression of the spots meaning, that expression indeed reached its asymptotic maximum value.

In the next step, these vector field properties were transformed into "vector"-trajectories using the R program velocito [31]. It summarizes vectors along the main directions pointing toward the major attractor-modules separated by "watersheds" (Figure 6B). The trajectories obtained resemble the gene-state trajectories connecting source and sink modules. However, RNA-velocity trajectories are more diverse, particularly, showing branched paths which point toward the local attractors of overexpression. Taken together, both SOM-based dynamic and RNA velocity analysis provide similar trajectories indicating that differentiation points toward overexpression modules of differentiated tissue.

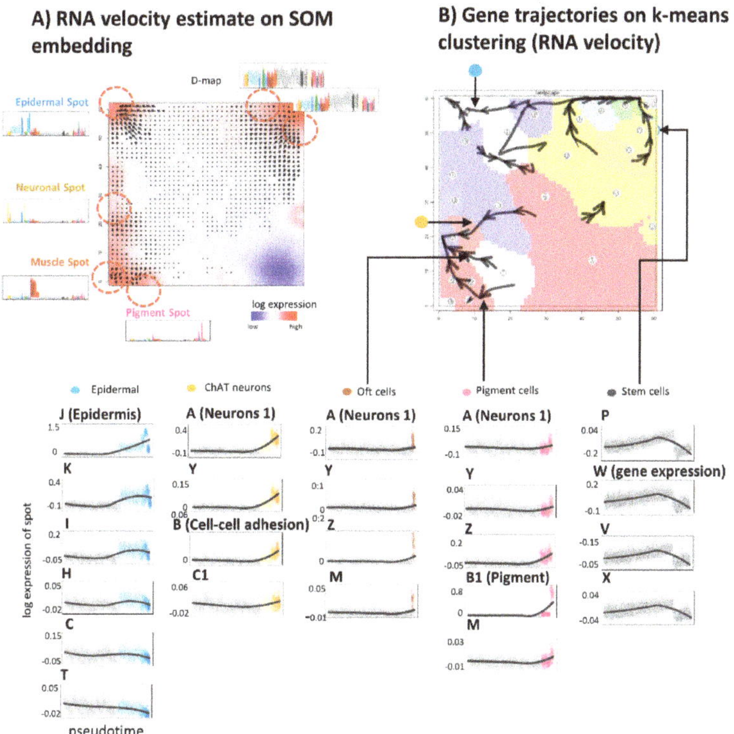

Figure 6. RNA velocity portrayal: (**A**) RNA velocity vector-field with metagene resolution identifies overexpression spot modules as attractors of overexpression except the "kernel" metagenes in the center of each spot (see text). (**B**) Velocity-trajectories using SOM area-segmentation indicate "watersheds" of differentiation between the attractors. The expression profiles illustrate expression changes along the selected trajectories.

4. Discussion

Single-cell RNA-omics opens novel opportunities for molecular profiling of thousands of individual cells in parallel. The analysis of cellular heterogeneity and of patterns across such molecular landscapes still faces challenges. Current computational approaches mostly share the following workflow components: (1) dimensionality reduction to extract most relevant axes of variability, (2) cluster and diversity analysis to extract most relevant groups of cells and similarity relations between them, and (3) feature selection to extract most relevant genes as markers characterizing the different cell types. Assumptions about temporal development enabled to perform diversity analysis of cross sectional data in such a way that one can deduce information about the (temporal) dynamics of cellular development and adaptation to changing environmental conditions. State-of-the-art methods accomplish this task in cell-state space, i.e., in the multidimensional manifold spanned by all possible states that cells can adopt upon development and changing environmental conditions. Hereby, i.e., by using scRNAseq data, the particular "state of the cell" is given by its transcriptional phenotype, i.e., the entirety of transcripts produced by a cell in a certain developmental and environmental situation and measured by the scRNAseq technology applied. We here "transpose" this view from the state space of the cells into the space of transcriptional patterns, which are derived from the scRNAseq data without linking gene expression directly to cell states.

From a methodical perspective, we complemented our SOM portrayal providing static transcriptomic landscapes by this dynamic aspect in order to deduce (pseudo-)temporal information from snapshot data obtained in cross sectional studies. Using SOM-portrayal we identified tissue-specific clusters of activated genes and linear trajectories linking the stemness-source and tissue-related sink clusters expressing transcriptional programs, which reflect the specific functional requirements of the respective tissue types. In addition, part of the sink spot-patterns reveal common activation of cellular functions in different tissues in support of our initial statement that cell- and gene-states are closely related but do not match in an one-to-one fashion. For example, morphogenesis-related genes (spot F) were activated in neuronal and muscle tissues as well, meaning that one and the same transcriptional program can contribute to different cell states. Our approach explicitly supports identification of such common gene states. Overall, the gene-state landscape decomposes into a directed graph describing development via linear and partly via branched paths. RNA velocity constitutes an independent approach to forecast future expression states. In the SOM landscape RNA velocity vectors point in the direction of the local slope of gene expression. This result seems reasonable and confirms this approach as an independent calculus for transcriptomic trajectories where the generated vector field provides the detailed shape of the local attractor fields pointing toward the overexpression spots.

We have chosen the Planarian example because it provides a multibranched lineage tree of tissue differentiation and also because for the first time it demonstrated the impact of single cell transcriptomics for establishing such lineage trees in a multicell organism [29]. Meanwhile a series of whole-organism and organ systems datasets have been generated, e.g., for Nematodes, Sea Anemones, Hydra, and Annelids, as well as the human hematopoietic system, lung, kidney, heart, gut endoderm, mesoderm, nervous system, and neural crest mostly including embryonic and adult tissue stadiums (see [16] and references cited therein). These transcriptomes contain information about multiple aspects of cell identity as provided by activated transcriptional programs (for example, cell cycle phases, metabolic states, cell-specific and tissue-specific molecular signatures). Hence, gene-state space-based "functional" views using SOM trajectories are expected to complement developmental paths in cell-state space in all these applications.

Novel developments further extend single cell technologies toward capturing other molecular features such as chromatin accessibility [46], DNA-methylomes, proteomes [47], and metabolomes [48], as well as multimodal measurements of them from the same single cells (e.g., mRNA and DNA [49]) and also in situ approaches complementing cell-intrinsic states with information on cell's local environment [50]. Hence, from an integrative perspective these diverse features reinforce an extended view of cell states as multidimensional vectors with diverse pheno- and genotypic features as components, which will enable holistic and comprehensive analysis of developmental trajectories in feature state space. Conceptually such a multidimensional continuous "landscape" was inspired by Waddington's epigenetic landscape [45]. Our gene-state space landscape can be interpreted as a gene-centric version of the Waddington-landscapes, which illustrates the development of cellular programs instead of cell types. Especially our difference construct between source and sink states generates SOM maps, which resemble the "down-hill along valleys" shape of the metaphoric epigenetic landscape in cell-state space.

We expect tumor progression as an important potential application of developmental trajectories where in addition to cell types (e.g., malignant cells, T cells, cancer associated fibroblasts) further diversity is expected to reflect distinct tumor stages and/or clonal lineages [51–53]. States in gene expression space are more dynamic (e.g., phases of cell cycle or adaptation of cells to different environment) than cell types. Tumor biology is typically driven by such subpopulations of cells. Their transcriptomic states activate along developmental lineages related to drug resistance, metastasis, immune evasion, and immunotherapy [54]. The role of genes along the respective state-space trajectories can be split into environmental stimuli accomplishing either core or transient functions according to [55]. RNA velocity vector fields possibly can be interpreted in terms of such transient effects modulating the local shape of the trajectories.

5. Conclusions

We here developed a method to "transpose" the concept of developmental trajectories in phase space of the cells into their phenotype space spanned by phenotype variables such as the expression of ten thousands of genes. After our proof-of-principle study using RNAseq data describing tissue development in the flatworm as example we see applications in different fields of biology and medicine, to other than transcriptomics data and also further needs for refinements of the method, e.g., to better disentangle the topology of the feature landscape and to link both, cell- and gene-state space trajectories.

Supplementary Materials: The following are available online at http://www.mdpi.com/2073-4425/11/10/1214/s1. Figure S1: Cellular markers and previous signatures of planarian single cell transcriptome, Figure S2: Positions of selected genes are shown in the map together with their expression profiles, Figure S3: Profiles and maps of tissue-wise differentially expressed genes and gene sets which were identified by [29], Table S1: References of selected marker genes of Figure S1, Table S2: Overview of module-wise enriched gene sets of the GO terms Biological Process, Cellular Component, Molecular function and Protein Domains.

Author Contributions: Conceived and wrote this paper: M.S., H.B.; performed analysis: M.S.; downstream analysis methods development: H.L.-W., M.S. All authors have read and agreed to the published version of the manuscript.

Funding: The author(s) acknowledge support from the German Research Foundation (DFG) and Universität Leipzig within the program of Open Access Publishing.

Conflicts of Interest: All other authors declare that they have no competing interests.

References

1. Kalisky, T.; Quake, S.R. Single-cell genomics. *Nat. Methods* **2011**, *8*, 311–314. [CrossRef] [PubMed]
2. Macaulay, I.C.; Voet, T. Single Cell Genomics: Advances and Future Perspectives. *PLoS Genet.* **2014**, *10*, e1004126. [CrossRef] [PubMed]
3. Bendall, S.C.; Davis, K.L.; Amir, E.A.D.; Tadmor, M.D.; Simonds, E.F.; Chen, T.J.; Shenfeld, D.K.; Nolan, G.P.; Pe'Er, D. Single-cell trajectory detection uncovers progression and regulatory coordination in human b cell development. *Cell* **2014**, *157*, 714–725. [CrossRef] [PubMed]
4. Haghverdi, L.; Büttner, M.; Wolf, F.A.; Buettner, F.; Theis, F.J. Diffusion pseudotime robustly reconstructs lineage branching. *Nat. Methods* **2016**, *13*, 845–848. [CrossRef]
5. Hanchate, N.K.; Kondoh, K.; Lu, Z.; Kuang, D.; Ye, X.; Qiu, X.; Pachter, L.; Trapnell, C.; Buck, L.B. Single-cell transcriptomics reveals receptor transformations during olfactory neurogenesis. *Science* **2015**, *350*, 1251–1255. [CrossRef]
6. Trapnell, C.; Cacchiarelli, D.; Grimsby, J.; Pokharel, P.; Li, S.; Morse, M.; Lennon, N.J.; Livak, K.J.; Mikkelsen, T.S.; Rinn, J.L. The dynamics and regulators of cell fate decisions are revealed by pseudotemporal ordering of single cells. *Nat. Biotechnol.* **2014**, *32*, 381–386. [CrossRef]
7. Campbell, K.R.; Yau, C. A descriptive marker gene approach to single-cell pseudotime inference. *Bioinformatics* **2019**, *35*, 28–35. [CrossRef]
8. Reid, J.E.; Wernisch, L. Pseudotime estimation: Deconfounding single cell time series. *Bioinformatics* **2016**, *32*, 2973–2980. [CrossRef] [PubMed]
9. Campbell, K.R.; Yau, C. Order Under Uncertainty: Robust Differential Expression Analysis Using Probabilistic Models for Pseudotime Inference. *PLoS Comput. Biol.* **2016**, *12*, e1005212. [CrossRef]
10. Saelens, W.; Cannoodt, R.; Todorov, H.; Saeys, Y. A comparison of single-cell trajectory inference methods. *Nat. Biotechnol.* **2019**, *37*, 547–554. [CrossRef]
11. Shin, J.; Berg, D.A.; Zhu, Y.; Shin, J.Y.; Song, J.; Bonaguidi, M.A.; Enikolopov, G.; Nauen, D.W.; Christian, K.M.; Ming, G.L.; et al. Single-Cell RNA-Seq with Waterfall Reveals Molecular Cascades underlying Adult Neurogenesis. *Cell Stem Cell* **2015**, *17*, 360–372. [CrossRef] [PubMed]
12. Street, K.; Risso, D.; Fletcher, R.B.; Das, D.; Ngai, J.; Yosef, N.; Purdom, E.; Dudoit, S. Slingshot: Cell lineage and pseudotime inference for single-cell transcriptomics. *BMC Genom.* **2018**, *19*, 477. [CrossRef] [PubMed]
13. Setty, M.; Tadmor, M.D.; Reich-Zeliger, S.; Angel, O.; Salame, T.M.; Kathail, P.; Choi, K.; Bendall, S.; Friedman, N.; Pe'Er, D. Wishbone identifies bifurcating developmental trajectories from single-cell data. *Nat. Biotechnol.* **2016**, *34*, 637–645. [CrossRef] [PubMed]

14. Qiu, X.; Mao, Q.; Tang, Y.; Wang, L.; Chawla, R.; Pliner, H.A.; Trapnell, C. Reversed graph embedding resolves complex single-cell trajectories. *Nat. Methods* **2017**, *14*, 979–982. [CrossRef] [PubMed]

15. Wolf, F.A.; Hamey, F.K.; Plass, M.; Solana, J.; Dahlin, J.S.; Göttgens, B.; Rajewsky, N.; Simon, L.; Theis, F.J. PAGA: Graph abstraction reconciles clustering with trajectory inference through a topology preserving map of single cells. *Genome Biol.* **2019**, *20*, 1–9. [CrossRef] [PubMed]

16. Wagner, D.E.; Klein, A.M. Lineage tracing meets single-cell omics: Opportunities and challenges. *Nat. Rev. Genet.* **2020**, *21*, 410–427. [CrossRef] [PubMed]

17. Orman, M.A.; Berthiaume, F.; Androulakis, I.P.; Ierapetritou, M.G. Advanced stoichiometric analysis of metabolic networks of mammalian systems. *Crit. Rev. Biomed. Eng.* **2011**, *39*, 511–534. [CrossRef]

18. Tritschler, S.; Büttner, M.; Fischer, D.S.; Lange, M.; Bergen, V.; Lickert, H.; Theis, F.J. Concepts and limitations for learning developmental trajectories from single cell genomics. *Development* **2019**, *146*, dev170506. [CrossRef]

19. Loeffler-Wirth, H.; Binder, H.; Willscher, E.; Gerber, T.; Kunz, M. Pseudotime dynamics in melanoma single-cell transcriptomes reveals different mechanisms of tumor progression. *Biology* **2018**, *7*, 23. [CrossRef]

20. Fan, J.; Slowikowski, K.; Zhang, F. Single-cell transcriptomics in cancer: Computational challenges and opportunities. *Exp. Mol. Med.* **2020**, *52*, 1452–1465. [CrossRef]

21. Valdes-Mora, F.; Handler, K.; Law, A.M.K.; Salomon, R.; Oakes, S.R.; Ormandy, C.J.; Gallego-Ortega, D. Single-cell transcriptomics in cancer immunobiology: The future of precision oncology. *Front. Immunol.* **2018**, *9*, 2582. [CrossRef] [PubMed]

22. Wirth, H.; Löffler, M.; Von Bergen, M.; Binder, H. Expression cartography of human tissues using self organizing maps. *BMC Bioinform.* **2011**, *12*, 306. [CrossRef] [PubMed]

23. Wirth, H.; von Bergen, M.; Binder, H. Mining SOM expression portraits: Feature selection and integrating concepts of molecular function. *BioData Min.* **2012**, *5*, 18. [CrossRef]

24. Kunz, M.; Löffler-Wirth, H.; Dannemann, M.; Willscher, E.; Doose, G.; Kelso, J.; Kottek, T.; Nickel, B.; Hopp, L.; Landsberg, J.; et al. RNA-seq analysis identifies different transcriptomic types and developmental trajectories of primary melanomas. *Oncogene* **2018**, *37*, 6136–6151. [CrossRef]

25. Gerber, T.; Willscher, E.; Loeffler-wirth, H.; Hopp, L.; Schartl, M.; Anderegg, U.; Camp, G.; Treutlein, B. Mapping heterogeneity in patient-derived melanoma cultures by. *Oncotarget* **2017**, *8*, 846–862. [CrossRef]

26. Camp, J.G.; Sekine, K.; Gerber, T.; Loeffler-Wirth, H.; Binder, H.; Gac, M.; Kanton, S.; Kageyama, J.; Damm, G.; Seehofer, D.; et al. Multilineage communication regulates human liver bud development from pluripotency. *Nature* **2017**, *546*, 533–538. [CrossRef] [PubMed]

27. Binder, H.; Löffler-Wirth, H. Analysis of large-scale OMIC data using Self Organizing Maps. *Encycl. Inf. Sci. Technol. Third Ed.* **2014**, 1642–1654. [CrossRef]

28. Binder, H.; Wirth, H.; Arakelyan, A.; Lembcke, K.; Tiys, E.S.; Ivanisenko, V.A.; Kolchanov, N.A.; Kononikhin, A.; Popov, I.; Nikolaev, E.N.; et al. Time-course human urine proteomics in space-flight simulation experiments. *BMC Genom.* **2014**, *15*, 1–19. [CrossRef]

29. Plass, M.; Solana, J.; Wolf, F.A.; Ayoub, S.; Misios, A.; Glažar, P.; Obermayer, B.; Theis, F.J.; Kocks, C.; Rajewsky, N. Supplementary Materials for Cell type atlas and lineage tree of a whole complex animal by single-cell transcriptomics. *Science* **2018**, *360*, eaaq1723. [CrossRef]

30. Huang, S.; Eichler, G.; Bar-Yam, Y.; Ingber, D.E. Cell fates as high-dimensional attractor states of a complex gene regulatory network. *Phys. Rev. Lett.* **2005**, *94*, 128701. [CrossRef]

31. La Manno, G.; Soldatov, R.; Zeisel, A.; Braun, E.; Hochgerner, H.; Petukhov, V.; Lidschreiber, K.; Kastriti, M.E.; Lönnerberg, P.; Furlan, A.; et al. RNA velocity of single cells. *Nature* **2018**, *560*, 494–498. [CrossRef] [PubMed]

32. Bergen, V.; Lange, M.; Peidli, S.; Wolf, F.A.; Theis, F.J. Generalizing RNA velocity to transient cell states through dynamical modeling. *Nat. Biotechnol.* **2020**, 1–7. [CrossRef] [PubMed]

33. Macosko, E.Z.; Basu, A.; Satija, R.; Nemesh, J.; Shekhar, K.; Goldman, M.; Tirosh, I.; Bialas, A.R.; Kamitaki, N.; Martersteck, E.M.; et al. Highly parallel genome-wide expression profiling of individual cells using nanoliter droplets. *Cell* **2015**, *161*, 1202–1214. [CrossRef] [PubMed]

34. Hayashi, T.; Asami, M.; Higuchi, S.; Shibata, N.; Agata, K. Isolation of planarian X-ray-sensitive stem cells by fluorescence-activated cell sorting. *Dev. Growth Differ.* **2006**, *48*, 371–380. [CrossRef]

35. Önal, P.; Grün, D.; Adamidi, C.; Rybak, A.; Solana, J.; Mastrobuoni, G.; Wang, Y.; Rahn, H.P.; Chen, W.; Kempa, S.; et al. Gene expression of pluripotency determinants is conserved between mammalian and planarian stem cells. *EMBO J.* **2012**, *31*, 2755–2769. [CrossRef] [PubMed]

36. Loeffler-Wirth, H.; Kreuz, M.; Hopp, L.; Arakelyan, A.; Haake, A.; Cogliatti, S.B.; Feller, A.C.; Hansmann, M.L.; Lenze, D.; Möller, P.; et al. A modular transcriptome map of mature B cell lymphomas. *Genome Med.* **2019**, *11*, 27. [CrossRef]

37. Loeffler-Wirth, H.; Kalcher, M.; Binder, H. OposSOM: R-package for high-dimensional portraying of genome-wide expression landscapes on bioconductor. *Bioinformatics* **2015**, *31*, 3225–3227. [CrossRef]

38. Hopp, L.; Loeffler-Wirth, H.; Nersisyan, L.; Arakelyan, A.; Binder, H. Footprints of Sepsis Framed Within Community Acquired Pneumonia in the Blood Transcriptome. *Front. Immunol.* **2018**, *9*, 1620. [CrossRef]

39. Van Der Maaten, L.; Hinton, G. Visualizing Data using t-SNE. *J. Mach. Learn Res.* **2008**, *9*, 2579–2605.

40. Farrell, J.A.; Wang, Y.; Riesenfeld, S.J.; Shekhar, K.; Regev, A.; Schier, A.F. Single-cell reconstruction of developmental trajectories during zebrafish embryogenesis. *Science* **2018**, *360*, eaar3131. [CrossRef]

41. Goldberg, A.V.; Tarjan, R.E. A New Approach to the Maximum-Flow Problem. *J. ACM* **1988**, *35*, 921–940. [CrossRef]

42. Wang, I.J. Topographic path analysis for modelling dispersal and functional connectivity: Calculating topographic distances using the topoDistance r package. *Methods Ecol. Evol.* **2020**, *11*, 265–272. [CrossRef]

43. Brandl, H.; Moon, H.K.; Vila-Farré, M.; Liu, S.Y.; Henry, I.; Rink, J.C. PlanMine—A mineable resource of planarian biology and biodiversity. *Nucleic Acids Res.* **2016**, *44*, D764–D773. [CrossRef] [PubMed]

44. Fincher, C.T.; Wurtzel, O.; de Hoog, T.; Kravarik, K.M.; Reddien, P.W. Cell type transcriptome atlas for the planarian Schmidtea mediterranea. *Science* **2018**, *360*, eaaq1736. [CrossRef]

45. Waddington, C.H. *The Strategy of the Genes*; George Allen & Unwin, Ltd: London, UK, 1957; ISBN 978-1-315-77932-4.

46. Lareau, C.A.; Duarte, F.M.; Chew, J.G.; Kartha, V.K.; Burkett, Z.D.; Kohlway, A.S.; Pokholok, D.; Aryee, M.J.; Steemers, F.J.; Lebofsky, R.; et al. Droplet-based combinatorial indexing for massive-scale single-cell chromatin accessibility. *Nat. Biotechnol.* **2019**, *37*, 916–924. [CrossRef]

47. Budnik, B.; Levy, E.; Harmange, G.; Slavov, N. SCoPE-MS: Mass spectrometry of single mammalian cells quantifies proteome heterogeneity during cell differentiation. *Genome Biol.* **2018**, *19*, 161. [CrossRef]

48. Duncan, K.D.; Fyrestam, J.; Lanekoff, I. Advances in mass spectrometry based single-cell metabolomics. *Analyst* **2019**, *144*, 782–793. [CrossRef]

49. Dey, S.S.; Kester, L.; Spanjaard, B.; Bienko, M.; Van Oudenaarden, A. Integrated genome and transcriptome sequencing of the same cell. *Nat. Biotechnol.* **2015**, *33*, 285–289. [CrossRef]

50. Lubeck, E.; Coskun, A.F.; Zhiyentayev, T.; Ahmad, M.; Cai, L. Single-cell in situ RNA profiling by sequential hybridization. *Nat. Methods* **2014**, *11*, 360–361. [CrossRef]

51. Sebastian, A.; Hum, N.R.; Martin, K.A.; Gilmore, S.F.; Peran, I.; Byers, S.W.; Wheeler, E.K.; Coleman, M.A.; Loots, G.G. Single-Cell Transcriptomic Analysis of Tumor-Derived Fibroblasts and Normal Tissue-Resident Fibroblasts Reveals Fibroblast Heterogeneity in Breast Cancer. *Cancers* **2020**, *12*, 1307. [CrossRef]

52. Qian, J.; Olbrecht, S.; Boeckx, B.; Vos, H.; Laoui, D.; Etlioglu, E.; Wauters, E.; Pomella, V.; Verbandt, S.; Busschaert, P.; et al. A pan-cancer blueprint of the heterogeneous tumor microenvironment revealed by single-cell profiling. *Cell Res.* **2020**, *30*, 745–762. [CrossRef] [PubMed]

53. Yu, K.; Hu, Y.; Wu, F.; Guo, Q.; Qian, Z.; Hu, W.; Chen, J.; Wang, K.; Fan, X.; Wu, X.; et al. Surveying brain tumor heterogeneity by single-cell RNA-sequencing of multi-sector biopsies. *Natl. Sci. Rev.* **2020**, *7*, 1306–1318. [CrossRef]

54. Tirosh, I.; Suvà, M.L. Deciphering Human Tumor Biology by Single-Cell Expression Profiling. *Annu. Rev. Cancer Biol.* **2019**, *3*, 151–166. [CrossRef]

55. Mar, J.C.; Quackenbush, J. Decomposition of Gene Expression State Space Trajectories. *PLoS Comput. Biol.* **2009**, *5*, e1000626. [CrossRef] [PubMed]

Publisher's Note: MDPI stays neutral with regard to jurisdictional claims in published maps and institutional affiliations.

 *genes*

Article

Novel Insights into the Protective Properties of ACTH$_{(4\text{-}7)}$PGP (Semax) Peptide at the Transcriptome Level Following Cerebral Ischaemia–Reperfusion in Rats

Ivan B. Filippenkov [1,*], Vasily V. Stavchansky [1], Alina E. Denisova [2], Vadim V. Yuzhakov [3], Larisa E. Sevan'kaeva [3], Olga Y. Sudarkina [1], Veronika G. Dmitrieva [1], Leonid V. Gubsky [2], Nikolai F. Myasoedov [1], Svetlana A. Limborska [1] and Lyudmila V. Dergunova [1]

[1]   Institute of Molecular Genetics, Russian Academy of Sciences, 123182 Moscow, Russia; bacbac@yandex.ru (V.V.S.); sudarolg@img.ras.ru (O.Y.S.); veronuska@mail.ru (V.G.D.); nfm@img.ras.ru (N.F.M.); limbor@img.ras.ru (S.A.L.); lvdergunova@mail.ru (L.V.D.)

[2]   Pirogov Russian National Research Medical University, Federal Center of Cerebrovascular Pathology and Stroke, Ministry of Health Care of Russian Federation, 117342 Moscow, Russia; dalina543@gmail.com (A.E.D.); gubskii@mail.ru (L.V.G.)

[3]   A. Tsyb Medical Radiological Research Center–Branch of the National Medical Research Radiological Center of the Ministry of Health of the Russian Federation, 249031 Obninsk, Russia; yuzh_vad@mail.ru (V.V.Y.); larisa.sevankaeva@mail.ru (L.E.S.)

*   Correspondence: filippenkov@img.ras.ru; Tel.: +7-499-196-1858

Received: 16 May 2020; Accepted: 18 June 2020; Published: 22 June 2020

Abstract: Cerebral ischaemia is the most common cause of impaired brain function. Biologically active peptides represent potential drugs for reducing the damage that occurs after ischaemia. The synthetic melanocortin derivative, ACTH$_{(4\text{-}7)}$PGP (Semax), has been used successfully in the treatment of patients with severe impairment of cerebral blood circulation. However, its molecular mechanisms of action within the brain are not yet fully understood. Previously, we used the transient middle cerebral artery occlusion (tMCAO) model to study the damaging effects of ischaemia–reperfusion on the brain transcriptome in rats. Here, using RNA-Seq analysis, we investigated the protective properties of the Semax peptide at the transcriptome level under tMCAO conditions. We have identified 394 differentially expressed genes (DEGs) (>1.5-fold change) in the brains of rats at 24 h after tMCAO treated with Semax relative to saline. Following tMCAO, we found that Semax suppressed the expression of genes related to inflammatory processes and activated the expression of genes related to neurotransmission. In contrast, ischaemia–reperfusion alone activated the expression of inflammation-related genes and suppressed the expression of neurotransmission-related genes. Therefore, the neuroprotective action of Semax may be associated with a compensation of mRNA expression patterns that are disrupted during ischaemia–reperfusion conditions.

Keywords: tMCAO; mRNA expression; RNA-Seq; synthetic melanocortin derivative ACTH$_{(4\text{-}7)}$PGP (Semax); peptide regulation

1. Introduction

Focal cerebral ischaemia or stroke is one of the leading causes of mortality and disability in developed countries. The development of effective strategies to treat ischaemic stroke is an important issue in modern medicine and pharmacology. A drug currently found to be efficient in cerebral stroke therapy is a synthetic neuroprotective peptide called Semax (Met-Glu-His-Phe-Pro-Gly-Pro). It bears a fragment of adrenocorticotropic hormone 4-7 (Met-Glu-His-Phe) and the C-terminal tripeptide Pro-Gly-Pro (PGP), last one was included to ensure the resistance of Semax to peptidases. The neuroprotective and neurotrophic effects of Semax are reliably established [1–8]. It has been shown that Semax improves cognitive function in both rodents and humans. The peptide facilitates acquisition of food-motivated and passive-avoidance tasks in healthy animals and has a protective effect in a model of stress-induced memory impairment in rats [9–11]. In experimental models of cerebral ischaemia Semax administration led to the recovery of the animals' ability to learn in a Morris water maze and passive-avoidance task [12,13]. A clinical study has shown the efficacy of Semax in the treatment of patients with ischaemic stroke. The peptide improves functional recovery and motor performance [14,15]. Nevertheless, the molecular mechanisms of its neuroprotective effects remain unclear.

Transcriptome analysis is one of the approaches used to study the functioning of the brain in ischaemia and under the action of drugs. To study the effect of drugs on the genes functioning under cerebral ischaemia conditions, models of cerebral ischaemia in laboratory animals have been widely used [16–22]. Previously, using models of global incomplete cerebral ischaemia we have shown the effect of Semax on the expression of the limited number of genes that encode neurotrophic factors and their receptors [23–25]. Using microchips RatRef-12 BeadChips, it was shown that peptide Semax affects the expression of genes related to the immune and vascular systems in rat fronto-parietal cortex after permanent middle cerebral artery occlusion (pMCAO) [26,27]. At present, to study the molecular mechanisms of the effects of Semax, we used another model of transient middle cerebral artery occlusion (tMCAO). This model reflects events that occur in ischaemic stroke in humans after treatment with thrombolytic agents [28,29]. The results of clinical studies indicate that, currently, thrombolysis is one of the most effective and affordable methods of treatment of ischaemic stroke [30,31]. In the tMCAO model conditions, which were based on endovascular artery occlusion (90 min) and subsequent reperfusion, we identified hundreds of differentially expressed genes (DEGs) [32]. In particular, we revealed the activation of a large number of genes involved in inflammation, the immune response, apoptosis and the stress response. Simultaneously, a massive downregulation of genes that ensure the functioning of neurotransmitter systems was observed in tMCAO conditions [32]. We studied genome-wide gene transcription of the subcortex tissue using the tMCAO model in an attempt to research the mechanisms of the neuroprotective effects of Semax. This study allowed us to observe alterations to the transcriptome profile and to reveal previously unknown compensation effects of Semax on the biological processes and signal pathways, which apparently provides the neuroprotective effects of the peptide in ischaemia–reperfusion (IR) conditions.

2. Materials and Methods

2.1. Animals

White 2-month-old male rats of the Wistar line (weight, 200–250 g) were obtained from the Experimental Radiology sector in A. Tsyb Medical Radiological Research Center, Obninsk, Russian Federation. The animals were maintained on a 12 h light/dark cycle at a temperature of 22–24 °C, with free access to food and water. The animals were divided into the "ischaemia–reperfusion" (IR) and "ischaemia–reperfusion after Semax administration" (IS) groups.

2.2. Transient Cerebral Ischaemia Rat Model

2.2.1. Operation

The transient cerebral ischaemia rat model was induced by endovascular occlusion of the right middle cerebral artery using a monofilament (Doccol Corporation, Sharon, MA, USA) for 90 min [33]. The rats were decapitated at 4.5 or 24 h after tMCAO (group IR—"IR4.5" or "IR24", respectively and group IS—"IS4.5" or "IS24", respectively). Prior to the surgical procedure, rats were anaesthetized using 3% isoflurane; the anaesthesia was maintained using 1.5–2% isoflurane and the EZ–7000 small animal anaesthesia system (E-Z Anaesthesia, Braintree, MA, USA). The sham-operated rats (SH) were subjected to a similar surgical procedure under anaesthesia (neck incision and separation of the bifurcation), but without tMCAO.

2.2.2. Semax Administration

To rats of the IS group, the neuropeptide Semax was administered intraperitoneally at a dose of 10 μg/100 g rat weight as previously described [11,26,27,34] at 90 min, as well as at 2.5 and 6.5 h after tMCAO. To rats of the IR groups, saline was administered intraperitoneally at 90 min, as well as at 2.5 and 6.5 h after the surgical procedure.

2.3. Magnetic Resonance Imaging

The magnetic resonance imaging (MRI) study of the characteristics and size of the ischaemic injury of rat brains was carried out using a small animal 7T system from ClinScan tomograph (Bruker BioSpin, Billerica, MA, USA). The standard protocol included the following modes: diffusion-weighted imaging (DWI) with mapping of the apparent diffusion coefficient (ADC) for assessing acute ischaemic damage (TR/TE = 9000/33 ms; b factors = 0 and 1000 s/mm^2; diffusion directions = 6; averages = 3; spectral fat saturation; FOV = 30 × 19.5 mm; slice thickness = 1.0 mm; matrix size = 86 × 56), and T2–weighted imaging (T2 WI) in the transverse plane (Turbo Spin Echo with restore magnetization pulse; turbo factor = 10; TR/TE = 5230/46 ms; averages = 2; spectral fat saturation; FOV = 30 × 21.1 mm; slice thickness = 0.7 mm; matrix size = 256 × 162). Three-dimensional time-of-flight magnetic resonance angiography (3D-TOF MRA) was used for visualization of the main arteries and control of the recanalization (3D Gradient Echo with RF spoiling and flow compensation; TR/TE = 30/4.55 ms; slabs = 4; flip angle = 70; averages = 1; FOV = 35 × 19.3 mm; slice thickness = 0.2 mm; matrix size = 320 × 176). A quantitative assessment of the volume of the infarction focus was performed using the ImageJ software package (Wayne Rasband, National Institute of Mental Health, Bethesda, MD, USA). MRI was performed immediately before decapitation in rats from the IR4.5 and IS4.5 groups. In rats from the IR24 and IS24 groups, MRI was performed twice: at 4.5 h after tMCAO and immediately before decapitation.

2.4. Histological Examination of Rat Brains

Tissue samples of rat brains at 24 h after occlusion ($n = 4$) and after sham operation ($n = 4$) were immersed in Bouin's fluid for 24 h and washed with 70% ethanol. The tissue samples were dehydrated and embedded in Histomix®(BioVitrum, St. Petersburg, Russia). Tissue sectioning was performed with the orientation of two tissue blocks for subsequent excision into coronary sections at the level from −4.0 to −0.5 and from −0.5 to +5 mm from the bregma. Sections with a thickness of 5–6 μm obtained through 0.5–1 mm on a microtome (Leica RM2235, Wetzlar, Germany) were stained with haematoxylin and eosin (H&E staining) and toluidine blue in Nissl modifications (BioVitrum, St. Petersburg, Russia) after dewaxing. Histological specimens were examined under a microscope (Leica DM 1000, Wetzlar, Germany) with a micrograph to digital camera (Leica ICC50 HD, Wetzlar, Germany). Morphological analysis was performed with allowance for normal and pathological central nervous system cells variants [35–37]. Stereotactic mapping of the damaged zones and accurate determination of the level of sections were performed according to an atlas of the rat brain [38].

2.5. RNA Isolation

Tissues were placed in RNAlater solution for 24 h at 0 °C and then stored at −70 °C. Total RNA from the subcortex, was isolated using TRIzol reagent (Invitrogen, Thermo Fisher Scientific, Waltham, MA, USA) and acid guanidinium thiocyanate-phenol-chloroform extraction [39]. The isolated RNA was treated with deoxyribonuclease I (DNase I) (Thermo Fisher Scientific) in the presence of RiboLock ribonuclease (RNase) inhibitor (Thermo Fisher Scientific), according to the manufacturer's recommended protocol. Deproteinization was performed using a 1:1 phenol:chloroform mixture.

The isolated RNA was precipitated with sodium acetate (3.0 M, pH 5.2) and ethanol. RNA integrity was checked using capillary electrophoresis (Experion, BioRad, Hercules, CA, USA). RNA integrity number (RIN) was at least 9.0.

2.6. RNA-Seq

Total RNA isolated from the subcortical structures of the brain, including the lesion focus, was used in this experiment. The RNA-Seq experiment was conducted with the participation of ZAO Genoanalytika, Russia. For RNA-Seq, the polyA fraction of the total RNA was obtained using the oligoT magnetic beads of the Dynabeads®mRNA Purification Kit (Ambion, Thermo Fisher Scientific, Waltham, MA, USA). cDNA (DNA complementary to RNA) libraries were prepared using the NEBNext®mRNA Library Prep Reagent Set (NEB, Ipswich, MA, USA). The concentration of cDNA libraries was measured using Qbit 2.0 and the Qubit dsDNA HS Assay Kit (Thermo Fisher Scientific, Waltham, MA, USA). The length distribution of library fragments was determined using the Agilent High Sensitivity DNA Kit (Agilent, Lexington, MA, USA). Sequencing was carried out using an Illumina HiSeq 1500 instrument. At least 10 million reads (1/50 nt) were generated.

2.7. RNA-Seq Data Analysis

Three pairwise comparisons of RNA-Seq results (IS4.5 versus IR4.5, IS24 versus IR24, and IS24 versus IS4.5) were used to analyse the action of the peptide Semax on the transcriptome. Each of the comparison groups (IS4.5, IS24, IR4.5 and IR24) included three animals ($n = 3$). According to the MRI data (DWI, T2 WI), the ischaemic foci in the brain of these animals had a subcortical localization. 3D-TOF MRA was used for visualization of the main arteries and also for the control of the recanalization. All genes were annotated on the NCBI Reference Sequence database. The levels of mRNA expression were measured as fragments per kilobase per million reads using the Cuffdiff program. Only genes that exhibited changes in expression >1.5-fold and had a p-values (t-test) adjusted using the Benjamini–Hochberg procedure lower 0.05 ($Pagj < 0.05$) were considered.

2.8. cDNA Synthesis

cDNA synthesis was conducted in 20 µL of reaction mixture containing 2 mg of RNA using the reagents of a RevertAid First Strand cDNA Synthesis Kit (Thermo Fisher Scientific, Waltham, MA, USA) in accordance with the manufacturer's instructions. Oligo $(dT)_{18}$ primers were used to analyse mRNA.

2.9. Real-Time Reverse Transcription Polymerase Chain Reaction (RT–PCR)

The 25 µL polymerase chain reaction (PCR) mixture contained 2 µL of 0.2× reverse transcriptase reaction sample, forward and reverse primers (5 pmol each), 5 µL of 5× reaction mixture (Evrogen Joint Stock Company, Moscow, Russia) including PCR buffer, Taq DNA polymerase, deoxyribonucleoside triphosphates (dNTP) and the intercalating dye SYBR Green I. Primers specific to the genes studied were selected using OLIGO Primer Analysis Software version 6.31 (Molecular Biology Insights, Inc., Cascade, CO, USA) and were synthesized by the Evrogen Joint Stock Company, Moscow, Russia (Supplementary Table S1). The amplification of cDNAs was performed using a StepOnePlus Real-Time PCR System (Applied Biosystems, Thermo Fisher Scientific, Waltham, Massachusetts, USA) in the following mode: stage 1 (denaturation), 95 °C, 10 min; stage 2 (amplification with fluorescence measured), 95 °C, 1 min; 65 °C, 1 min; 72 °C, 1 min (40 cycles).

2.10. Data Analysis of Real-Time RT–PCR and Statistics

Two reference genes *Gapdh* and *Rpl3* were used to normalize the cDNA samples [40]. Calculations were performed using BestKeeper, version 1 (gene-quantification, Freising-Weihenstephan, Bavaria, Germany) [41] and Relative Expression Software Tool (REST) 2005 software (gene-quantification, Freising-Weihenstephan, Bavaria, Germany) [42]. The manual at the site 'REST.-gene-quantification.info' was used to evaluate expression target genes relative to the expression levels of the reference genes. The values were calculated as $Ef^{Ct(ref)}/Ef^{Ct(tar)}$, where Ef is the PCR efficiency, Ct(tar) is the average threshold cycle (Ct) of the target gene, Ct(ref) is the average Ct of the reference gene, and $Ef^{Ct(ref)}$ is the geometric average Ef^{Ct} of the reference genes. PCR efficiencies were assessed using the amplification of a series of standard dilutions of cDNAs and computed using REST software [42]. The efficiency values for all PCR reactions were in the range 1.83 to 2.08 (Supplementary Table S1). At least five animals were included in each comparison group (n ≥ 5). When comparing data groups, statistically significant differences were considered with the probability $P < 0.05$. Additional calculations were performed using Microsoft Excel (Microsoft Office 2010, Microsoft, Redmond, WA, USA).

2.11. Functional Analysis

Database for Annotation, Visualization and Integrated Discovery (DAVID v6.8, Laboratory of Human Retrovirology and Immunoinformatics, Frederick, MD, USA) [43], Gene Set Enrichment Analysis (GSEA) [44], gProfileR [45] and The PANTHER database (Protein ANalysis THrough Evolutionary Relationships) [46] were used to annotate the functions of the differentially expressed genes. Only functional annotations that had a *p*-values adjusted using the Benjamini–Hochberg procedure lower 0.05 (*Pagj* < 0.05) were considered. Hierarchical cluster analysis of DEGs was performed using Heatmapper (Wishart Research Group, University of Alberta, Ottawa, Canada) [47]. Volcano plot were constructed by Microsoft Excel (Microsoft Office 2010, Microsoft, Redmond, WA, USA).

2.12. Availability of Data and Material

RNA-sequencing data have been deposited in the Sequence Read Archive database under accession code SRP148632 (SAMN09235828-SAMN09235839) [48], and PRJNA491404 (SAMN10077190-SAMN10077195) [49].

2.13. Ethics Approval and Consent to Participate

All manipulations with experimental animals were approved by the Animal Care Committee of the Pirogov Russian National Research Medical University (Approved ID: 15-2015, 2 November 2015) and were carried out in accordance with the Directive 2010/63/EU of the European Parliament and the Council of European Union on the protection of animals used for scientific purposes issued on 22 September 2010.

3. Results

3.1. Characterization of tMCAO Model Conditions Using MRI

The location of ischaemic foci was detected in animals under tMCAO conditions using the T2-weighted imaging (T2 WI) and diffusion-weighted imaging (DWI) of magnetic resonance imaging (MRI). According to MRI, animals after tMCAO under the influence of saline and Semax had the ischaemic zone that was localized in the subcortical structures of the brain from the side of occlusion or spread to the cortex (Supplementary Table S2). Figure S1 shows the MRI of ischaemic foci with a subcortical localization at 4.5 and 24 h after tMCAO under the influence of saline or Semax and MRI of rat brain after sham operation (Supplementary Figure S1).

3.2. Histopathological Characterization of Rat Brain

Twenty-four hours after tMCAO, no pathomorphological changes were observed in the cortex of the cerebral hemispheres or the medial region of the subcortical nuclei (Figure 1a,b). After Nissl staining there were no changes in the cytoarchitectonics of the layers of neurons on the cortex of the cerebral hemispheres (Figure 1e). Groups of morphologically unchanged neurons with a basophilic substance in their cytoplasm were located in the medial caudoputamen zone (Figure 1f). However, in the lateral areas of the subcortical region of the right hemisphere, the extensive ischaemic damage to brain tissue was noted as oedema and reticular foci of pale stained foci (Figure 1a,e). In the histological specimens stained with haematoxylin and eosin (H&E), in the caudorostral range from −2.0 to −0.3 mm from the bregma, the ischaemic stroke formation zone captured most of the striatum to the outer capsule with a clear visualization of the nucleus infarction and had an elongated shape in the dorsoventral direction.

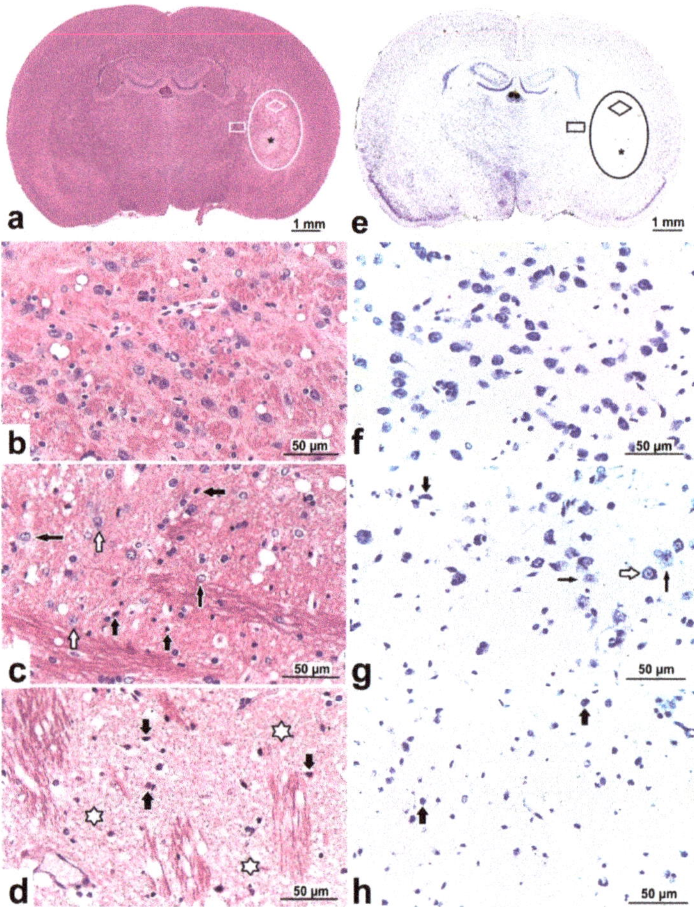

Figure 1. Photomicrographs of haematoxylin and eosin-stained (**a–d**) and Nissl-stained (**e–h**) sections of the rat brain in tMCAO model conditions. (**a,e**) Serial coronal rat brain sections at the level of −2.0 mm from the bregma. Rectangles indicate the normal tissue in the medial region of the caudoputamen. The oval indicates the damaged area involving the lateral region of the caudoputamen nucleus of the right hemisphere. Asterisks indicate necrotic tissue in the central core of an infarct. (**b,f**) High-magnification images of the normal tissue in the areas of panels (**a,e**) marked with a rectangle. (**c,g**) Areas of panels (**a,e**) marked with a rhomb. Hypoxic damage to neurons, with pyknotic nuclei and pericellular oedema indicated in the ischaemic zone (thick black arrows); decrease of nuclear basophilia and Nissl substance in the neurons (thin black arrows); intact neurons (white arrows). (**d,h**) Areas of panels (**a,e**) marked with asterisks. Ischaemic necrosis of the brain tissue in the central core of an infarct; destruction of the neuropil (white asterisks); dead "pyknotic" neurons (thick black arrows), Nissl substance disappearing.

On microscopic examination of the penumbra, we found narrowing of the capillary lumen, pronounced perivascular oedema, sparsity and weak staining of the neuropil due to sponginess and vacuolization, and the appearance of numerous hyperchromic neurons with with pyknotic nuclei and pericellular oedema (Figure 1c). A significant part of the neurons was in a state of hypoxic damage or death. In addition, in the ischaemic zones there was a decrease in the content of Nissl-positive neurons, both with partial and total chromatolysis (Figure 1g). Only a few neurons did not have obvious pathological changes.

The nucleus of the infarction was represented by a complete loss of nervous tissue with destruction of pyknotic neurons and all elements of the neuropil (Figure 1d), with Nissl substance disappearing in their perikaryon (Figure 1h).

In the contralateral hemisphere of ischaemic rats (Figure 1), and in both hemisphere of sham-operated rats (Supplementary Figure S2), the histological pattern of the capillary network and the morphology of neurons in the brain cortex, subcortical nuclei and the intermediate brain corresponded to the norm.

3.3. RNA-Seq Analysis of the Effect of Semax on the Transcriptome after tMCAO

Using RNA-seq we assessed the effect of Semax on the mRNA level of genes functioning in the subcortical structures of the rat brain at 4.5 and 24 h after tMCAO. Animals subjected to ischaemia–reperfusion (IR) that received saline were used as controls for animals after Semax administration in a tMCAO model at 4.5 h (IS4.5 versus IR4.5) and 24 h after operation (IS24 versus IR24), respectively. From more than 17,000 genes, no significant DEGs (>1.5-fold) were seen after Semax administration versus saline in IR conditions at 4.5 h after tMCAO, but we identified 394 (191 up- and 203 downregulated) DEGs under the influence of Semax at 24 h after tMCAO (IS24 versus IR24) (Figure 2a). The volcano plot shows the up- and downregulated DEGs in IS24 versus IR24 groups (Figure 2b).

We used the real-time RT–PCR analysis of the expression of 4 up- (*Cplx2, Gabra5, Neurod6, Ptk2b*), 4 down- (*Hspb1, Hspa1(a,b), Fos, Lrg1*) and 2 non-significantly (*Ttr, Vegfa*) regulated genes to verify the RNA-Seq results (Supplementary Figure S3). The real-time RT–PCR results confirmed the RNA-Seq data adequately. The differences in the methodology and statistical processing of data used in each case obviously could contribute to some differences in the level of mRNA expression identified by these methods.

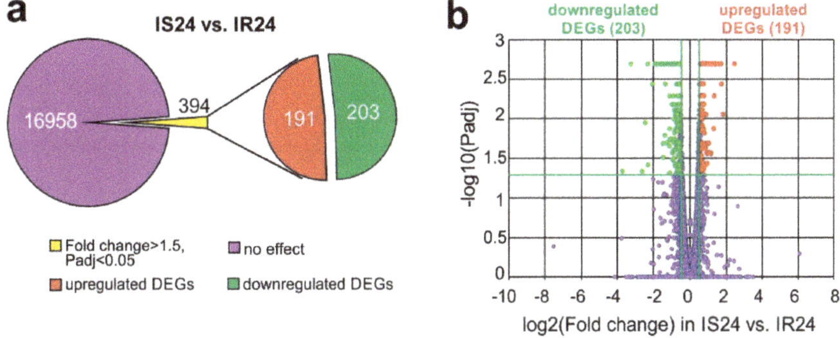

Figure 2. RNA-Seq analysis of the effect of Semax on the transcriptome at 24 h after tMCAO. (**a**) RNA-Seq results presented are for in IS24 versus IR24. The numbers in the diagram sectors indicate the number of DEGs. (**b**) Volcano plots show the distributions of genes between the IS24 and IR24 groups. Up- and downregulated DEGs are represented as red and green dots, respectively (fold change > 1.50. *Padj* < 0.05). Not differential expressed genes are represented as dark purple dots (fold change ≤ 1.50. *Padj* ≥ 0.05).

3.4. Functional Annotation of Semax-Induced DEGs Identified in the Rat Brain Subcortex at 24 h after tMCAO

Using the PANTHER tool (version 15.0, Free Software Foundation, Inc., Boston, MA, USA) we classified Semax-induced DEGs according to the molecular functions of their encoded proteins. In IS24 versus IR24, upregulated DEGs were predominantly associated with receptor activity, signal transducer activity and transporter activity, whereas downregulated DEGs were associated with structural molecule activity, and binding and catalytic activity (Supplementary Table S3).

Using the DAVID program, the functional categories of the proteins encoded by the DEGs were identified (Supplementary Table S4). The total number of them associated with the effects of Semax at 24 h after tMCAO (IS24 versus IR24) was 34. The top 5 functional categories with the smallest *Padj* were Disulfide bond, Glycoprotein, Signal, Secreted and Calcium. Functional categories of neurotransmission systems of cells (Synapse, Postsynaptic cell membrane, Calmodulin-binding, and Cell junction) were associated predominantly with upregulated DEGs in IS24 versus IR24 (Supplementary Table S4). Simultaneously, functional categories of immunity and inflammatory responses (Immunity, Intermediate filament, and Innate immunity) were associated predominantly with downregulated DEGs in IS24 versus IR24 (Supplementary Table S4).

3.5. Differences in Rat Brain Transcriptomes Following Ischaemia and after Semax Administration

Previously, using RNA-Seq, we identified 1939 DEGs (1109 up- and 839 downregulated) under IR conditions after tMCAO at 24 h versus sham operation at 24 h after surgical procedure (IR24 versus SH24) [32]. In this study, we identified 313 DEGs that overlapped for IS24 versus IR24 and IR24 versus SH24 (Figure 3a, Supplementary Table S5). Venn diagrams with only up-regulated DEGs and only down-regulated DEGs in both conditions are shown in Figure 3b,c. We found only the *Mx1* gene, which encodes an interferon-induced GTP-binding protein was upregulated in both IS24 versus IR24 and IR24 versus SH24, simultaneously (Figure 3b), and did not found DEGs which were downregulated in both conditions (Figure 3c). It should be noted that Semax initiated mRNA expression that counteracted the effects of IR injury. Hierarchical cluster analysis shows the Semax increases the expression levels of genes that reduce expression by the action of ischaemia and vice versa, that is, it compensates for the effect of ischaemia (Figure 3d). In particular, we found 155 up- and 157 downregulated DEGs in IS24 versus IR24 that counteracted IR in IR24 versus SH24 (Supplementary Table S5). The top 10 genes with the greatest fold change in IS24 versus IR24 are presented in Figure 3e. The figure shows that the expression levels of genes *Gpr6*, *Neu2*, *Hes5*, *Gpr88*, and *Drd2* was increased by up to 2.5-fold in IS24 versus IR24 and was decreased by up to 4-fold in IR24 versus SH24. Simultaneously, the expression of genes *Glycam1*, *S100a9*, *Ccl6*, *Gh1*, and *Hspa1(a,b)* was decreased by 3-fold in IS24 versus IR24 and was increased by up to 5-fold in IR24 versus SH24.

Moreover, 81 genes (35 up- and 46 downregulated) altered their expression in IS24 versus IR24 but did not alter it in IR24 versus SH24 (Figure 3a, Supplementary Table S6). These DEGs predominantly have catalytic and binding activity, but additionally have receptor and signal transducer (*Gabrb1*, *Grin2b*, *Folr1*, *Bdnf*), as well as transporter (*Kcnj13*, *Slc1a2*, *Slco1a5*) activity (Supplementary Table S7).

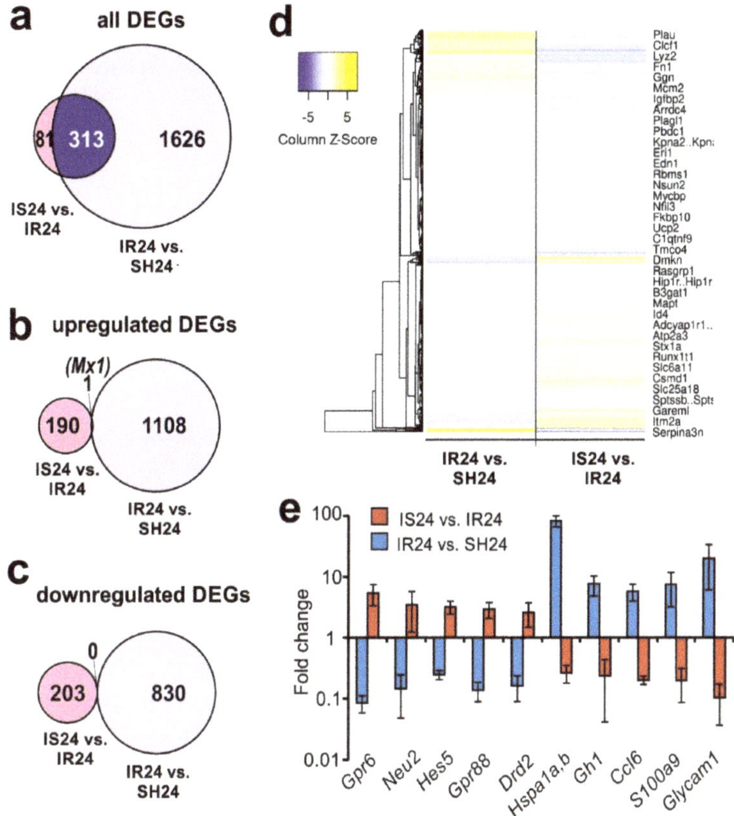

Figure 3. Comparison of the results obtained in two pairwise comparisons of IS24 versus IR24 and IR24 versus SH24. (**a**–**c**) Venn diagrams of DEGs in two pairwise comparisons of IS24 versus IR24 and IR24 versus SH24. Comparison for all (**a**), up- (**b**) and downregulated (**c**) DEGs. (**d**) Hierarchical cluster analysis of all DEGs in IS24 versus IR24 and IR24 versus SH24. Each column represents a comparison group, and each row represents a DEG. Yellow strips represent high relative expressions and blue strips represent low relative expressions, $n = 3$ per group. (**e**) The top ten genes that exhibited the greatest fold change in expression in IS24 versus IR24. The data are presented as the mean ± standard error of the mean. The cut-off of gene-expression changes was 1.50. Only those genes, whose *Padj* < 0.05 were selected for analysis.

3.6. Signalling Pathways Associated with Semax-Induced DEGs under tMCAO Model Conditions

Semax-induced DEGs identified in the rat brain subcortex under tMCAO model conditions were annotated according to a Kyoto Encyclopedia of Genes and Genomes (KEGG) using DAVID v6.8. We found 25 signalling pathways associated with DEGs in IS24 versus IR24. Previously, we found 82 signalling pathways associated with DEGs detected in the brains of rats in the IR24 versus SH24 groups [32]. Among them, 17 signalling pathways overlapped between IS24 versus IR24 and IR24 versus SH24 (Figure 4a). These signalling pathways are involved in neurotransmission, drug metabolism and inflammation. There were 8 signalling pathways (Porphyrin and chlorophyll metabolism, Drug metabolism—other enzymes, Phagosome, PI3K-Akt and MAPK and other signalling pathways) predominantly associated with the upregulated DEGs in IR24 versus SH24 and downregulated DEGs in IS24 versus IR24 (Figure 4b). Conversely, there were 8 signalling pathways (Amphetamine addiction, Retrograde endocannabinoid signalling, Glutamatergic and Dopaminergic synapses and other signalling pathways) predominantly associated with the downregulated DEGs in IR24 versus SH24 and upregulated DEGs in IS24 versus IR24. Additionally, similar results were obtained using GSEA (KEGG, Reactome, BioCarta, PID) and gProfileR (KEGG, Reactome, WikiPathways). So, based on GSEA data, upregulated DEGs in IS24 versus IR24 were associated with Neuronal System, Signaling by G-protein-coupled receptors (GPCR), Transmission across Chemical Synapses, etc; whereas downregulated DEGs were associated with Innate Immune System, Neutrophil degranulation, Cytokine Signalling in Immune system, etc (Supplementary Table S8). Also, based on gProfileR data, upregulated DEGs in IS24 versus IR24 were associated with, Calcium signaling pathway, Dopaminergic, Cholinergic and Glutamatergic synapse, etc; whereas downregulated DEGs were associated with Phagosome, IL-17, TNF, p53 signaling pathway, etc (Supplementary Table S9). Thus, Semax initiated neurotransmitter and inflammatory response that counteracted IR at 24 h after tMCAO.

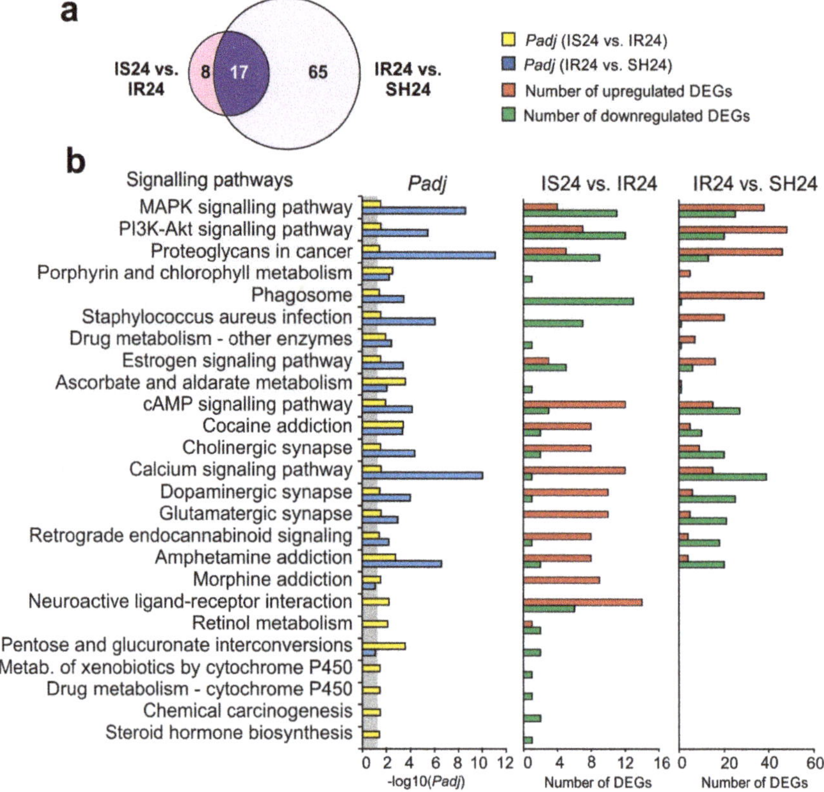

Figure 4. Analysis of the signalling pathways associated with DEGs at 24 h after tMCAO. (**a**) The numbers of signalling pathways overlapped in two pairwise comparisons: IS24 versus IR24 and IR24 versus SH24. (**b**) KEGG database analyses of DEGs in two pairwise comparisons IS24 versus IR24 and IR24 versus SH24 was carried out according to the DAVID database. The number of upregulated and downregulated DEGs, as well as the *p*-values adjusted using the Benjamini–Hochberg procedure (*Padj*), are shown. Only those genes and signalling pathways whose *Padj* < 0.05 were selected for analysis. *Padj* ≥ 0.05 are enclosed in the gray background.

4. Discussion

Drugs, based on native regulatory peptides, are used to treat various pathological conditions, including peptide drugs that help restore brain function after acute cerebrovascular accidents. In studies to investigate the molecular mechanisms of action of such neuroprotective peptides, experimental models of ischaemia in animals are of great importance. For example, the neuroprotective effects of orexin (OxA) [50] and poly-arginine R18 and NA-1 peptides [51], as well as the anti-inflammatory and anti-oxidant effects of cordymin peptide [52] were identified in studies employing middle cerebral artery occlusion-induced focal cerebral ischaemia in rats. Semax, a nootropic neuropeptide, has been used in neurological practice for many years in the treatment of acute and chronic disorders, including ischaemic stroke and its consequences [14,15,53]. However, its molecular mechanisms of action are not yet fully understood. There are many examples of the use of transcriptome analysis to study the mechanisms of action of a number of peptides including PACAP38 [54], Sal-like 4 peptide [55] and OxA [50]. The first genome-wide analysis of Semax action on the rat brain transcriptome was conducted using a permanent MCAO model that was induced by direct permanent electrical coagulation of the distal segment of the left middle cerebral artery [26,27]. The genome-wide biochip data analysis detected DEGs associated with several biological processes and signalling pathways. In the first hours after pMCAO, a significant increase in the expression of transcription factor genes was observed in the presence of Semax. These genes encode proteins that trigger signalling to correct the destructive processes associated with ischaemic conditions. Semax also stimulated the expression of genes that encode growth factors involved in trophic and protective processes, as well as the vascularization of damaged tissues. Additionally, Semax administration had a significant effect on the expression of genes associated with the immune response [26,27].

In the current study of the protective mechanisms of action of the Semax peptide, we aimed to identify the DEGs, as well as the biological processes and signalling pathways involved in the response of rat brain cells to Semax action under ischaemia–reperfusion (IR) conditions. Therefore, we used the tMCAO model, which involves occlusion of the MCA and subsequent restoration of blood flow [22,56]. Importantly, the tMCAO model in rats replicates circumstances that arise in the human brain after ischaemic stroke treated with thrombolytic medications [28,29]. MRI (Supplementary Table S2 and Figure S1) and histopathological analysis (Figure 1) detected the ischaemic focus and the penumbra region in the subcortical structures of the brain after tMCAO. A study of the molecular mechanisms of cell death under pMCAO and tMCAO conditions, conducted by Ford et al. [57] revealed molecular functions and biological processes unique to each model. Genes uniquely altered by tMCAO included a number of genes related to inflammatory and oxidative stress, whereas pMCAO led to the induction of genes that were more associated with metabolic activity and cellular signalling [57]. In addition, reperfusion after ischaemia leads to additional injury, including disturbance of the blood-brain barrier via destruction of the endothelial microvascular brain cells, and damage of brain cells through the accumulation of excess oxygen radicals and apoptosis [22,58–61].

Using RNA-Seq to determine the mRNA expression profile in the subcortical structures of rat brain in response to Semax administration, a significant effect of Semax was identified at 24 h after tMCAO. We identified 394 DEGs (>1.5-fold change) following Semax administration at 24 h after tMCAO, but there were no DEGs identified following Semax administration at 4.5 h after tMCAO. An analysis of the expression of 10 genes by real-time RT-PCR confirmed the RNA-Seq results. A large number of genes, altered by Semax in the tMCAO model, encode transcripts and proteins involved in cellular processes, metabolic processes, biological regulation, responses to external stimuli and multicellular organism-associated processes.

In a previous study, we revealed the large transcriptome response of cells in the subcortical structures of rat brain at 24 h after tMCAO and identified the biological processes and signalling pathways involved in the response to IR damage [32]. These conditions produced activation of a large number of genes involved in inflammation, the immune response, apoptosis, stress responses, ribosome function, DNA replication and other processes (e.g., *Hspa1(a,b)*, *Hspb1*, *Lrg1*, *Jun*, *Socs3*, *Cish*, *Cd14*, *Cd63*, *Cd74*, *Ccl6*, *Ccl9*, *Nfkb2*, *Fos* and others). In contrast, the expression of many genes related to neurotransmitter system function was inhibited (e.g., *Chrm1*, *Chrm4*, *Cplx2*, *Drd1*, *Drd2*, *Gabra5*, *Gria3*, *Grm3*, *Grm5*, *Gpr6*, *Gpr88*, *Htr6*, *Neurod6* and others) [32].

In the current study, we first compared the transcriptome response of brain cells to IR (IR24 versus SH24) and the effect of Semax treatment under IR conditions (IS24 versus IR24). These analyses revealed that the effect of the peptide on several hundred genes was opposite to that of IR damage. Under IR conditions, Semax administration upregulated the expression of *Gpr6, Neu2, Hes5, Gpr88, Drd2* and others, the genes predominantly involved in the neuronal receptor actions and neurogenesis, and downregulated the expression of chemokine genes (*Ccl6, Ccl9*), early response genes (*Hspa1(a,b), Fos, Jun*) and others (Figure 3e). These data suggest that Semax can normalize the expression of many genes disrupted during IR.

The functional annotation of Semax-induced DEGs under tMCAO model conditions also demonstrated the compensatory effect of the peptide on the functioning of a number of signalling and metabolic systems after IR. Semax suppressed the inflammatory, and activated the neurotransmitter signalling pathways, at 24 h after tMCAO (Figure 4), whereas we previously revealed that IR activated inflammatory, but suppressed neurotransmitter signalling pathways at 24 h after tMCAO [32]. Therefore, Semax initiated genetic responses that counteracted those produced by IR injury.

Our data on the activation of neurotransmitter signalling pathways by Semax are in good agreement with previous research, which reported a positive modulatory effect of Semax on the striatal serotonergic system and the ability of Semax to enhance the striatal release of dopamine [62]. Semax has also been shown to modulate GABA- and glycine-activated ionic currents in isolated cerebral neurons [63]. Earlier it was shown that Semax (1 μM) did not affect cyclic adenosine monophosphate (cAMP) levels in HEK293 cells expressing the human melanocortin 4 receptor (MCR4), but antagonized the cAMP-inducing effect of the melanocortin derivative α-melanocyte-stimulating hormone (α-MSH) in this cells [64]. Additionally, in models of brain ischaemia and inflammation, α-MSH exhibits strong anti-inflammatory, neurogenic and neuroprotective effects [65–67]. Therefore, Semax may recapitulate some important neuroprotective actions of the melanocortins.

The major limitation of our study is the lack of data confirming the change in the protein level of mRNA expression of detected DEGs, and any functional evidence supporting the change in the mRNA expression of these genes. In the future, further studies on the functioning of a limited number of key genes of the detected signaling pathways at the RNA and protein levels are necessary to be done. However, the main achievement is the full transcriptome study with the identification of specific signaling pathways involved in the molecular mechanisms of the peptide action. Bulk RNA-seq, in combination with bioinformatic approaches, remain dominant and valuable tools for the simultaneous analysis of changes in mRNA expression of a huge number of genes and the search for signaling pathways that are involved in the response to specific exposure. It was precisely the analysis of RNA-seq that allowed us to establish that the Semax peptide suppresses the expression of genes associated with inflammatory processes and activates the expression of genes related to neurotransmission. Significantly, these transcriptome changes driven by Semax are the opposite of those caused by IR.

In summary, our data indicate that an important feature of the neuroprotective action of Semax is the normalization of mRNA expression patterns that are disrupted during IR conditions, particularly those associated with anti-inflammatory processes and activation of neurotransmitter systems.

5. Conclusions

In conclusion, the study of the transcriptome profile of cells in the subcortical structures of the brain with administration of the neuropeptide drug Semax under tMCAO conditions led to the identification of DEGs that encode proteins that participate in various functional categories, biological processes and signalling pathways, via which brain cells form a response to ischaemia–reperfusion (IR). Under tMCAO conditions, we found that Semax initiated mRNA expression that counteracted IR. In particular, Semax suppressed inflammatory and activated neurotransmitter genes, whereas the genetic response initiated by IR activates inflammatory and suppresses neurotransmitter genes. We revealed significant compensation effects of Semax peptide on inflammatory and neurotransmitter genetic responses after tMCAO, which may account for the neuroprotective action of Semax under IR conditions. Thus, an important feature of Semax is the normalization of mRNA expression patterns that are disturbed during ischaemia.

Supplementary Materials: The following are available online at http://www.mdpi.com/2073-4425/11/6/681/s1, Figure S1: Characterization of tMCAO model conditions using MRI, Figure S2: Morphology of brain tissues after sham operation, Figure S3: Real-time reverse transcription polymerase chain reaction (RT-PCR) verification of the RNA-Seq results, Table S1: The characterization of the primers for real-time RT-PCR, Table S2. Quantitative assessment of the volume of the ischemic brain injury at 24 hours after tMCAO using magnetic resonance imaging, Table S3: Analysis of the molecular function associated with Semax-induced DEGs (IS24 versus IR24) using the PANTHER tool, Table S4: Analysis of the functional categories of the proteins encoded by DEGs in IS24 versus IR24, Table S5: Analysis of DEGs that overlapped between IS24 versus IR24 and IR24 versus SH24, Table S6: Analysis of DEGs that changed their expression in IS24 versus IR24, but did not change it in IR24 versus SH24, Table S7: Analysis of the molecular functions associated with the DEGs using the PANTHER tool, Table S8: Analysis of the functional annotations of DEGs in IS24 versus IR24 using GSEA, Table S9: Analysis of the functional annotations of DEGs in IS24 versus IR24 using gProfileR.

Author Contributions: Conceptualization, I.B.F., S.A.L. and L.V.D.; methodology, I.B.F. and L.V.D.; software, I.B.F. and O.Y.S.; validation, I.B.F., V.V.S. and V.G.D.; formal analysis, I.B.F. and V.V.S.; investigation, I.B.F., V.V.S, A.E.D. and L.E.S.; resources, I.B.F.; data curation, S.A.L, N.F.M., V.V.Y. and L.V.G.; writing—original draft preparation, I.B.F. and A.E.D.; writing—review and editing, I.B.F., S.A.L. and L.V.D.; visualization, I.B.F.; supervision, S.A.L. and L.V.D.; project administration, S.A.L. and N.F.M.; funding acquisition, S.A.L. All authors have read and agreed to the published version of the manuscript.

Funding: This research was funded by Russian Science Foundation (RSF), grant 19-14-00268.

Acknowledgments: The authors thank Vladimir P. Chekhonin, MD, PhD, Pirogov Russian National Research Medical University, Serbsky Federal Medical Research Centre of Psychiatry and Narcology, Moscow, Russia for help in organizing an MRI study. The authors thank Ekaterina V. Medvedeva from Institute for Regenerative Medicine, Sechenov First Moscow State Medical University for the help in conducting surgeries.

Conflicts of Interest: The authors declare no conflict of interest.

References

1. Asmarin, I.P.; Nezavibat'ko, V.N.; Miasoedov, N.F.; Kamenskiĭ, A.A.; Grivennikov, I.A.; Ponomareva-Stepnaia, M.A.; Andreeva, L.A.; Kaplan, A.I.; Koshelev, V.B.; Riasina, T.V. A nootropic adrenocorticotropin analog 4-10-semax (15 years experience in its design and study). *Zh. Vyssh. Nerv. Deiat. Im. I P Pavlova* **1997**, *47*, 420–430. [PubMed]

2. de Wied, D. Neuropeptides in learning and memory processes. *Behav. Brain Res.* **1997**, *83*, 83–90. [CrossRef]

3. Kaplan, A.I.; Koshelev, V.B.; Nezavibat'ko, V.N.; Ashmarin, I.P. [Increased resistance to hypoxia effected by the neuropeptide preparation SEMAX]. *Fiziol. Cheloveka* **1992**, *18*, 104–107. [PubMed]

4. Storozhevykh, T.P.; Tukhbatova, G.R.; Senilova, Y.E.; Pinelis, V.G.; Andreeva, L.A.; Myasoyedov, N.F. Effects of semax and its Pro-Gly-Pro fragment on calcium homeostasis of neurons and their survival under conditions of glutamate toxicity. *Bull. Exp. Biol. Med.* **2007**, *143*, 601–604. [CrossRef]

5. Inozemtsev, A.N.; Bokieva, S.B.; Karpukhina, O.V.; Gumargalieva, K.Z.; Kamensky, A.A.; Myasoedov, N.F. Semax prevents learning and memory inhibition by heavy metals. *Dokl. Biol. Sci. Proc. Acad. Sci. USSR Biol. Sci. Sect.* **2016**, *468*, 112–114. [CrossRef]

6. Bashkatova, V.G.; Koshelev, V.B.; Fadyukova, O.E.; Alexeev, A.A.; Vanin, A.F.; Rayevsky, K.S.; Ashmarin, I.P.; Armstrong, D.M. Novel synthetic analogue of ACTH 4-10 (Semax) but not glycine prevents the enhanced nitric oxide generation in cerebral cortex of rats with incomplete global ischemia. *Brain Res.* **2001**, *894*, 145–149. [CrossRef]

7. Gusev, E.; Skvortsova, V. *Brain Ischaemia*, 1st ed.; Meditsina Publishers: Moscow, Russia, 2001; p. 328.

8. Tabbì, G.; Magrì, A.; Giuffrida, A.; Lanza, V.; Pappalardo, G.; Naletova, I.; Nicoletti, V.G.; Attanasio, F.; Rizzarelli, E. Semax, an ACTH4-10 peptide analog with high affinity for copper(II) ion and protective ability against metal induced cell toxicity. *J. Inorg. Biochem.* **2015**, *142*, 39–46. [CrossRef]

9. Glazova, N.Y.; Sebentsova, E.A.; Manchenko, D.M.; Andreeva, L.A.; Dergunova, L.V.; Levitskaya, N.G.; Limborska, S.A.; Myasoedov, N.F. The Protective Effect of Semax in a Model of Stress-Induced Impairment of Memory and Behavior in White Rats. *Biol. Bull.* **2018**, *45*, 394–399. [CrossRef]

10. Levitskaya, N.G.; Glazova, N.Y.; Sebentsova, E.A.; Manchenko, D.M.; Vilensky, D.A.; Andreeva, L.A.; Kamensky, A.A.; Myasoedov, N.F. Investigation of the Spectrum of Physiological Activities of the Heptapeptide Semax, an ACTH 4–10 Analogue. *Neurochem. J.* **2008**, *2*, 95–101.

11. Ashmarin, I.; Nezavibatko, V.; Levitskaya, N.; Koshelev, V.; Kamensky, A. Design and Investigation of an ACTH(4-10) Analog Lacking D-Amino Acids and Hydrophobic Radicals. *Neurosci. Res. Commun.* **1995**, *16*, 105–112.

12. Silachev, D.N.; Shram, S.I.; Shakova, F.M.; Romanova, G.A.; Myasoedov, N.F. Formation of spatial memory in rats with ischemic lesions to the prefrontal cortex; effects of a synthetic analog of ACTH(4-7). *Neurosci. Behav. Physiol.* **2009**, *39*, 749–756. [CrossRef]

13. Romanova, G.A.; Silachev, D.N.; Shakova, F.M.; Kvashennikova, Y.N.; Viktorov, I.V.; Shram, S.I.; Myasoedov, N.F. Neuroprotective and antiamnesic effects of Semax during experimental ischemic infarction of the cerebral cortex. *Bull. Exp. Biol. Med.* **2006**, *142*, 663–666. [CrossRef]

14. Gusev, E.I.; Martynov, M.Y.; Kostenko, E.V.; Petrova, L.V.; Bobyreva, S.N. [The efficacy of semax in the tretament of patients at different stages of ischemic stroke]. *Zhurnal Nevrol. i psikhiatrii Im. S.S. Korsakova* **2018**, *118*, 61–68. [CrossRef] [PubMed]

15. Gusev, E.I.; Skvortsova, V.I.; Miasoedov, N.F.; Nezavibat'ko, V.N.; Zhuravleva, E.I.; Vanichkin, A.V. [Effectiveness of semax in acute period of hemispheric ischemic stroke (a clinical and electrophysiological study)]. *Zhurnal Nevrol. i psikhiatrii Im. S.S. Korsakova* **1997**, *97*, 26–34.

16. Hernandes, M.S.; Lassègue, B.; Hilenski, L.L.; Adams, J.; Gao, N.; Kuan, C.-Y.; Sun, Y.-Y.; Cheng, L.; Kikuchi, D.S.; Yepes, M.; et al. Polymerase delta-interacting protein 2 deficiency protects against blood-brain barrier permeability in the ischemic brain. *J. Neuroinflamm.* **2018**, *15*, 45. [CrossRef]

17. Wang, L.; Liu, H.; Zhang, L.; Wang, G.; Zhang, M.; Yu, Y. Neuroprotection of Dexmedetomidine against Cerebral Ischemia-Reperfusion Injury in Rats: Involved in Inhibition of NF-κB and Inflammation Response. *Biomol. Ther. (Seoul)* **2017**, *25*, 383–389. [CrossRef] [PubMed]

18. Ghazavi, H.; Hoseini, S.J.; Ebrahimzadeh-Bideskan, A.; Mashkani, B.; Mehri, S.; Ghorbani, A.; Sadri, K.; Mahdipour, E.; Ghasemi, F.; Forouzanfar, F.; et al. Fibroblast Growth Factor Type 1 (FGF1)-Overexpressed Adipose-Derived Mesenchaymal Stem Cells (AD-MSCFGF1) Induce Neuroprotection and Functional Recovery in a Rat Stroke Model. *Stem Cell Rev.* **2017**, *13*, 670–685. [CrossRef]

19. Kim, Y.; Kim, Y.S.; Kim, H.Y.; Noh, M.-Y.; Kim, J.Y.; Lee, Y.-J.; Kim, J.; Park, J.; Kim, S.H. Early Treatment with Poly(ADP-Ribose) Polymerase-1 Inhibitor (JPI-289) Reduces Infarct Volume and Improves Long-Term Behavior in an Animal Model of Ischemic Stroke. *Mol. Neurobiol.* **2018**, *55*, 7153–7163. [CrossRef]

20. Ismael, S.; Zhao, L.; Nasoohi, S.; Ishrat, T. Inhibition of the NLRP3-inflammasome as a potential approach for neuroprotection after stroke. *Sci. Rep.* **2018**, *8*, 5971. [CrossRef]

21. Berger, C.; Stauder, A.; Xia, F.; Sommer, C.; Schwab, S. Neuroprotection and glutamate attenuation by acetylsalicylic acid in temporary but not in permanent cerebral ischemia. *Exp. Neurol.* **2008**, *210*, 543–548. [CrossRef]

22. White, B.C.; Sullivan, J.M.; DeGracia, D.J.; O'Neil, B.J.; Neumar, R.W.; Grossman, L.I.; Rafols, J.A.; Krause, G.S. Brain ischemia and reperfusion: Molecular mechanisms of neuronal injury. *J. Neurol. Sci.* **2000**, *179*, 1–33. [CrossRef]

23. Medvedeva, E.V.; Dmitrieva, V.G.; Povarova, O.V.; Limborska, S.A.; Skvortsova, V.I.; Myasoedov, N.F.; Dergunova, L.V. Effect of semax and its C-terminal fragment Pro-Gly-Pro on the expression of VEGF family genes and their receptors in experimental focal ischemia of the rat brain. *J. Mol. Neurosci.* **2013**, *49*, 328–333. [CrossRef] [PubMed]

24. Dmitrieva, V.G.; Povarova, O.V.; Skvortsova, V.I.; Limborska, S.A.; Myasoedov, N.F.; Dergunova, L.V. Semax and Pro-Gly-Pro activate the transcription of neurotrophins and their receptor genes after cerebral ischemia. *Cell. Mol. Neurobiol.* **2010**, *30*, 71–79. [CrossRef] [PubMed]

25. Stavchansky, V.V.; Yuzhakov, V.V.; Botsina, A.Y.; Skvortsova, V.I.; Bondurko, L.N.; Tsyganova, M.G.; Limborska, S.A.; Myasoedov, N.F.; Dergunova, L.V. The effect of Semax and its C-end peptide PGP on the morphology and proliferative activity of rat brain cells during experimental ischemia: A pilot study. *J. Mol. Neurosci.* **2011**, *45*, 177–185. [CrossRef] [PubMed]

26. Medvedeva, E.V.; Dmitrieva, V.G.; Povarova, O.V.; Limborska, S.A.; Skvortsova, V.I.; Myasoedov, N.F.; Dergunova, L.V. The peptide semax affects the expression of genes related to the immune and vascular systems in rat brain focal ischemia: Genome-wide transcriptional analysis. *BMC Genomics* **2014**, *15*, 228. [CrossRef]

27. Medvedeva, E.V.; Dmitrieva, V.G.; Limborska, S.A.; Myasoedov, N.F.; Dergunova, L.V. Semax, an analog of ACTH(4-7), regulates expression of immune response genes during ischemic brain injury in rats. *Mol. Genet. Genomics* **2017**, *292*, 635–653. [CrossRef]

28. Nieswandt, B.; Kleinschnitz, C.; Stoll, G. Ischaemic stroke: A thrombo-inflammatory disease? *J. Physiol.* **2011**, *589*, 4115–4123. [CrossRef]

29. Ryang, Y.-M.; Dang, J.; Kipp, M.; Petersen, K.-U.; Fahlenkamp, A.V.; Gempt, J.; Wesp, D.; Rossaint, R.; Beyer, C.; Coburn, M. Solulin reduces infarct volume and regulates gene-expression in transient middle cerebral artery occlusion in rats. *BMC Neurosci.* **2011**, *12*, 113. [CrossRef]

30. Lopes, R.D.; Piccini, J.P.; Hylek, E.M.; Granger, C.B.; Alexander, J.H. Antithrombotic therapy in atrial fibrillation: Guidelines translated for the clinician. *J. Thromb. Thrombolysis* **2008**, *26*, 167–174. [CrossRef]

31. Alexandrov, A.V.; Hall, C.E.; Labiche, L.A.; Wojner, A.W.; Grotta, J.C. Ischemic stunning of the brain: Early recanalization without immediate clinical improvement in acute ischemic stroke. *Stroke* **2004**, *35*, 449–452. [CrossRef]

32. Dergunova, L.V.; Filippenkov, I.B.; Stavchansky, V.V.; Denisova, A.E.; Yuzhakov, V.V.; Mozerov, S.A.; Gubsky, L.V.; Limborska, S.A. Genome-wide transcriptome analysis using RNA-Seq reveals a large number of differentially expressed genes in a transient MCAO rat model. *BMC Genomics* **2018**, *19*, 655. [CrossRef] [PubMed]

33. Koizumi, J.; Yoshida, Y.; Nakazawa, T.; Ooneda, G. Experimental studies of ischemic brain edema. *Nosotchu* **1986**, *8*, 1–8. [CrossRef]

34. Miasoedova, N.F.; Skvortsova, V.I.; Nasonov, E.L.; Zhuravleva, E.I.; Grivennikov, I.A.; Arsen'eva, E.L.; Sukhanov, I.I. [Investigation of mechanisms of neuro-protective effect of semax in acute period of ischemic stroke]. *Zhurnal Nevrol. i psikhiatrii Im. S.S. Korsakova* **1999**, *99*, 15–19.

35. Li, Y.; Powers, C.; Jiang, N.; Chopp, M. Intact, injured, necrotic and apoptotic cells after focal cerebral ischemia in the rat. *J. Neurol. Sci.* **1998**, *156*, 119–132. [CrossRef]

36. Lipton, P. Ischemic cell death in brain neurons. *Physiol. Rev.* **1999**, *79*, 1431–1568. [CrossRef] [PubMed]

37. Garcia, J.H.; Liu, K.F.; Ho, K.L. Neuronal necrosis after middle cerebral artery occlusion in Wistar rats progresses at different time intervals in the caudoputamen and the cortex. *Stroke* **1995**, *26*, 636–642, discussion 643. [CrossRef] [PubMed]

38. BUTTNER-ENNEVER, J. The Rat Brain in Stereotaxic Coordinates, 3rd edn. By George Paxinos and Charles Watson. (Pp. xxxiii+80; illustrated; f$69.95 paperback; ISBN 0 12 547623; comes with CD-ROM.) San Diego: Academic Press. 1996. *J. Anat.* **1997**, *191*, 315–317. [CrossRef]

39. Chomczynski, P.; Sacchi, N. Single-step method of RNA isolation by acid guanidinium thiocyanate-phenol-chloroform extraction. *Anal. Biochem.* **1987**, *162*, 156–159. [CrossRef]

40. Bustin, S.A.; Benes, V.; Garson, J.A.; Hellemans, J.; Huggett, J.; Kubista, M.; Mueller, R.; Nolan, T.; Pfaffl, M.W.; Shipley, G.L.; et al. The MIQE guidelines: Minimum information for publication of quantitative real-time PCR experiments. *Clin. Chem.* **2009**, *55*, 611–622. [CrossRef]

41. Pfaffl, M.W.; Horgan, G.W.; Dempfle, L. Relative expression software tool (REST) for group-wise comparison and statistical analysis of relative expression results in real-time PCR. *Nucleic Acids Res.* **2002**, *30*, e36. [CrossRef]

42. Pfaffl, M.W.; Tichopad, A.; Prgomet, C.; Neuvians, T.P. Determination of stable housekeeping genes, differentially regulated target genes and sample integrity: BestKeeper–Excel-based tool using pair-wise correlations. *Biotechnol. Lett.* **2004**, *26*, 509–515. [CrossRef] [PubMed]

43. Huang, D.W.; Sherman, B.T.; Lempicki, R.A. Systematic and integrative analysis of large gene lists using DAVID bioinformatics resources. *Nat. Protoc.* **2009**, *4*, 44–57. [CrossRef] [PubMed]

44. Subramanian, A.; Tamayo, P.; Mootha, V.K.; Mukherjee, S.; Ebert, B.L.; Gillette, M.A.; Paulovich, A.; Pomeroy, S.L.; Golub, T.R.; Lander, E.S.; et al. Gene set enrichment analysis: A knowledge-based approach for interpreting genome-wide expression profiles. *Proc. Natl. Acad. Sci. USA* **2005**, *102*, 15545–15550. [CrossRef] [PubMed]

45. Reimand, J.; Arak, T.; Adler, P.; Kolberg, L.; Reisberg, S.; Peterson, H.; Vilo, J. g:Profiler-a web server for functional interpretation of gene lists (2016 update). *Nucleic Acids Res.* **2016**, *44*, 83–89. [CrossRef]

46. Mi, H.; Huang, X.; Muruganujan, A.; Tang, H.; Mills, C.; Kang, D.; Thomas, P.D. PANTHER version 11: Expanded annotation data from Gene Ontology and Reactome pathways, and data analysis tool enhancements. *Nucleic Acids Res.* **2017**, *45*, 183–189. [CrossRef]

47. Babicki, S.; Arndt, D.; Marcu, A.; Liang, Y.; Grant, J.R.; Maciejewski, A.; Wishart, D.S. Heatmapper: Web-enabled heat mapping for all. *Nucleic Acids Res.* **2016**, *44*, W147–W153. [CrossRef]

48. National Center for Biotechnology Information. Available online: https://www.ncbi.nlm.nih.gov/Traces/study/?acc=SRP148632 (accessed on 10 May 2020).

49. National Center for Biotechnology Information. Available online: https://www.ncbi.nlm.nih.gov/sra/PRJNA491404 (accessed on 10 May 2020).

50. Wang, C.-M.; Pan, Y.-Y.; Liu, M.-H.; Cheng, B.-H.; Bai, B.; Chen, J. RNA-seq expression profiling of rat MCAO model following reperfusion Orexin-A. *Oncotarget* **2017**, *8*, 113066–113081. [CrossRef]

51. Milani, D.; Cross, J.L.; Anderton, R.S.; Blacker, D.J.; Knuckey, N.W.; Meloni, B.P. Neuroprotective efficacy of poly-arginine R18 and NA-1 (TAT-NR2B9c) peptides following transient middle cerebral artery occlusion in the rat. *Neurosci. Res.* **2017**, *114*, 9–15. [CrossRef]

52. Wang, J.; Liu, Y.-M.; Cao, W.; Yao, K.-W.; Liu, Z.-Q.; Guo, J.-Y. Anti-inflammation and antioxidant effect of Cordymin, a peptide purified from the medicinal mushroom Cordyceps sinensis, in middle cerebral artery occlusion-induced focal cerebral ischemia in rats. *Metab. Brain Dis.* **2012**, *27*, 159–165. [CrossRef]

53. Gusev, E.I.; Skvortsova, V.I.; Chukanova, E.I. [Semax in prevention of disease progress and development of exacerbations in patients with cerebrovascular insufficiency]. *Zhurnal Nevrol. i psikhiatrii Im. S.S. Korsakova* **2005**, *105*, 35–40.

54. Hori, M.; Nakamachi, T.; Shibato, J.; Rakwal, R.; Shioda, S.; Numazawa, S. Unraveling the Specific Ischemic Core and Penumbra Transcriptome in the Permanent Middle Cerebral Artery Occlusion Mouse Model Brain Treated with the Neuropeptide PACAP38. *Microarrays* **2015**, *4*, 2–24. [CrossRef] [PubMed]

55. Liu, B.H.; Jobichen, C.; Chia, C.S.B.; Chan, T.H.M.; Tang, J.P.; Chung, T.X.Y.; Li, J.; Poulsen, A.; Hung, A.W.; Koh-Stenta, X.; et al. Targeting cancer addiction for SALL4 by shifting its transcriptome with a pharmacologic peptide. *Proc. Natl. Acad. Sci. USA* **2018**, *115*, 7119–7128. [CrossRef] [PubMed]

56. Canazza, A.; Minati, L.; Boffano, C.; Parati, E.; Binks, S. Experimental models of brain ischemia: A review of techniques, magnetic resonance imaging, and investigational cell-based therapies. *Front. Neurol.* **2014**, *5*, 19. [CrossRef] [PubMed]

57. Ford, G.; Xu, Z.; Gates, A.; Jiang, J.; Ford, B.D. Expression Analysis Systematic Explorer (EASE) analysis reveals differential gene expression in permanent and transient focal stroke rat models. *Brain Res.* **2006**, *1071*, 226–236. [CrossRef] [PubMed]

58. Rosenberg, G.A.; Estrada, E.Y.; Dencoff, J.E. Matrix metalloproteinases and TIMPs are associated with blood-brain barrier opening after reperfusion in rat brain. *Stroke* **1998**, *29*, 2189–2195. [CrossRef] [PubMed]

59. Lochhead, J.J.; McCaffrey, G.; Quigley, C.E.; Finch, J.; DeMarco, K.M.; Nametz, N.; Davis, T.P. Oxidative stress increases blood-brain barrier permeability and induces alterations in occludin during hypoxia-reoxygenation. *J. Cereb. Blood Flow Metab.* **2010**, *30*, 1625–1636. [CrossRef] [PubMed]

60. Ritter, L.S.; Orozco, J.A.; Coull, B.M.; McDonagh, P.F.; Rosenblum, W.I. Leukocyte accumulation and hemodynamic changes in the cerebral microcirculation during early reperfusion after stroke. *Stroke* **2000**, *31*, 1153–1161. [CrossRef]

61. Nour, M.; Scalzo, F.; Liebeskind, D.S. Ischemia-reperfusion injury in stroke. *Interv. Neurol.* **2013**, *1*, 185–199. [CrossRef]

62. Eremin, K.O.; Kudrin, V.S.; Saransaari, P.; Oja, S.S.; Grivennikov, I.A.; Myasoedov, N.F.; Rayevsky, K.S. Semax, an ACTH(4-10) analogue with nootropic properties, activates dopaminergic and serotoninergic brain systems in rodents. *Neurochem. Res.* **2005**, *30*, 1493–1500. [CrossRef]

63. Sharonova, I.N.; Bukanova, Y.V.; Myasoedov, N.F.; Skrebitskii, V.G. Modulation of GABA- and Glycine-Activated Ionic Currents with Semax in Isolated Cerebral Neurons. *Bull. Exp. Biol. Med.* **2018**, *164*, 612–616. [CrossRef]

64. Adan, R.A.H.; Oosterom, J.; Ludvigsdottir, G.; Brakkee, J.H.; Burbach, J.P.H.; Gispen, W.H. Identification of antagonists for melanocortin MC3, MC4 and MC5 receptors. *Eur. J. Pharmacol. Mol. Pharmacol.* **1994**, *269*, 331–337. [CrossRef]

65. Giuliani, D.; Ottani, A.; Neri, L.; Zaffe, D.; Grieco, P.; Jochem, J.; Cavallini, G.M.; Catania, A.; Guarini, S. Multiple beneficial effects of melanocortin MC4 receptor agonists in experimental neurodegenerative disorders: Therapeutic perspectives. *Prog. Neurobiol.* **2017**, *148*, 40–56. [CrossRef] [PubMed]

66. Lisak, R.P.; Benjamins, J.A. Melanocortins, Melanocortin Receptors and Multiple Sclerosis. *Brain Sci.* **2017**, *7*, 104. [CrossRef] [PubMed]

67. Mykicki, N.; Herrmann, A.M.; Schwab, N.; Deenen, R.; Sparwasser, T.; Limmer, A.; Wachsmuth, L.; Klotz, L.; Köhrer, K.; Faber, C.; et al. Melanocortin-1 receptor activation is neuroprotective in mouse models of neuroinflammatory disease. *Sci. Transl. Med.* **2016**, *8*, 362ra146. [CrossRef]

Review

The Metabolization Profile of the *CYP2D6* Gene in Amerindian Populations: A Review

Luciana P. C. Leitão [1], Tatiane P. Souza [1], Juliana C. G. Rodrigues [1], Marianne R. Fernandes [1], Sidney Santos [1,2] and Ney P. C. Santos [1,2,*]

[1] Oncology Research Center, Federal University of Pará, Belém, Pará 66073, Brazil; colaresluciana@gmail.com (L.P.C.L.); xtatixsouza@gmail.com (T.P.S.); julianacgrodrigues@gmail.com (J.C.G.R.); fernandesmr@yahoo.com.br (M.R.F.); sidneysantos@ufpa.br (S.S.)

[2] Laboratory of Human and Medical Genetics, Institute of Biological Science, Federal University of Pará, Belém, Pará 66077-830, Brazil

* Correspondence: npcsantos.ufpa@gmail.com

Received: 9 January 2020; Accepted: 14 February 2020; Published: 28 February 2020

Abstract: Background: the *CYP2D6* gene is clinically important and is known to have a number of variants. This gene has four distinct metabolization profiles that are determined by the different allelic forms present in the individual. The relative frequency of these profiles varies considerably among human populations around the world. Populations from more isolated regions, such as Native Americans, are still relatively poorly studied, however. Even so, recent advances in genotyping techniques and increasing interest in the study of these populations has led to a progressive increase in publication rates. Given this, the review presented here compiled the principal papers published on the *CYP2D6* gene in Amerindian populations to determine the metabolic profile of this group. Methods: a systematic literature review was conducted in three scientific publication platforms (Google Scholar, Science Direct, and Pubmed). The search was run using the keywords "CYP2D6 Amerindians" and "CYP2D6 native Americans". Results: a total of 13 original papers met the inclusion criteria established for this study. All the papers presented frequencies of the different *CYP2D6* alleles in Amerindian populations. Seven of the papers focused specifically on Amerindian populations from Mexico, while the others included populations from Argentina, Chile, Costa Rica, Mexico, Paraguay, Peru, and the United States. The results of the papers reviewed here showed that the extensive metabolization profile was the most prevalent in all Amerindian populations studied to date, followed by the intermediate, slow, and ultra-rapid, in that order. Conclusion: the metabolization profiles of the Amerindian populations reviewed in the present study do not diverge in any major way from those of other populations from around the world. Given the paucity of the data available on Amerindian populations, further research is required to better characterize the metabolization profile of these populations to ensure the development of adequate therapeutic strategies.

Keywords: Amerindians; *CYP2D6*; phenotype; genetic polymorphisms; metabolization profile; Native Americans

---

1. Introduction

1.1. Rationale

The cytochrome P450 2D6 (*CYP2D6*) is a member of the cytochrome P450 gene family, a group of enzymes that is responsible for phase I metabolism and the elimination of a variety of endogenous substrates and a diverse array of drugs [1]. The *CYP2D6* gene is the most frequently studied member of the P450 gene family in clinical research [2]. While this enzyme represents only a small proportion (1.3–4.3%) of all hepatic Cytochrome P450 enzymes (CYPs), it is known to metabolize more than 20%

of all the drugs processed in the human liver, including at least 160 therapeutic targets, including antidepressants, antipsychotics, antiarrhythmics, opioid analgesics, anticancer agents, and other drug classes [3].

The *CYP2D6* gene is in the Chr22q13.1 region, close to two non-functional pseudogenes (*CYP2D7* and *CYP2D8*), and has a vast number of polymorphisms [4]. Up to now, over 125 allelic variants of the *CYP2D6* gene have been documented (PHARMVAR-https://www.pharmvar.org/gene/CYP2D6). These variants modify enzyme activity in various ways, that can be classified in four phenotypic groups: poor metabolizers (PM), intermediate metabolizers (IM), extensive metabolizers (EM), and ultrarapid metabolizers (UM) [5]. These differences in enzyme activity may result in both inter-individual and interethnic variation, with the relationship between the CYP2D6 genotypes and phenotypes being of considerable importance for the determination of therapeutic strategies in clinical practice [6].

The ethnic profile of a population may play an important role in the differentiation of the drug metabolism capacity among its individuals. Around the world, different populations carry alleles that characterize distinct *CYP2D6* phenotypes that vary among ethnic groups and, in turn, geographic regions. Llerena et al. (2014) reviewed the allelic variability of *CYP2D6* in major geographic regions and discovered that *CYP2D6*4* (an allele with inactive enzyme activity, which is present in PM phenotypes) is most frequent in Europe [7]. Alleles associated with decreased enzyme activity are frequent in Asia and East Asia (*CYP2D6**10), Africa and Afrodescedant populations (*CYP2D6**17 and *29), with *CYP2D6**41 and amplifications being found in Middle Eastern populations [7]. Zhou et al. (2017) analyzed the data available on the 1000 genomes platform, and offered considerable variability in several members of the CYP450 family, in particular *CYP2D6*, which varied greatly among the different populations for which data were available on the platform [8].

Although *CYP450* family genotypes and metabolic phenotypes have been studied extensively in different parts of the world [9,10], few data are available for some populations, such as those of the Native Americans (or Amerindians). The 2010 United States (US) Census [11] recorded a population of approximately 6.6 million Native Americans in this country. Worldwide, hispanics (including those from Latin America, the Caribbean, and the US) comprise a total population of more than 600 million individuals (http://data.worldbank.org/region/LAC), the equivalent of 8.4% of the world's population. In addition, approximately 45 million Amerindians live in Latin America, representing 8.3% of the total population of this region (https://www.cepal.org/en/infografias/los-pueblos-indigenas-en-america-latina).

Present-day Latin American populations are the end result of a process that began with migrations from northeastern Asia around 15,000–18,000 years ago, and was finalized over the past five centuries, following the arrival of Europeans and Africans, which led to extensive admixture [12,13]. Most New World populations reflect some degree of this process of admixture. The Mexican-American population is a multiple admixture of different ethnic groups, combining the genetic background of a number of Native American peoples, derived mainly from a single migration of Asians through Beringia, with white Europeans from Spain. There are approximately 70 groups of Amerindians in Mexico, with more than 85 languages and dialects, located mainly in the center and southeastern portions of the country, with an estimated total population of 10,113,411 [14]. Native Chileans, an important Amerindian population in South America, make up around 9% of the total population of Chile, that is 1,585,680 individuals who self-identified as Amerindian, according to Casen (2015) [15]. The Mapuche live in southern South America, in both Chile and Argentina, an area divided by the Andes, and account for approximately 84% of native Chileans [15,16].

In 2011, the population of Costa Rica was 4,301,712. This population is the result of admixture initiated during colonial times, which has blended Mesoamerican and European (primarily Spanish) genes with those from sub-Saharan African [17]. Approximately 2% of this population self-identified as Amerindian, distributed in different provinces all around the country, with distinct dialects and customs, including some individuals that do not belong to a tribe, but live in rural or urban areas [18]. In 2013, Paraguay had 19 indigenous tribes belonging to five linguistic families, with a total population of 112,848 individuals. In this country, indigenous peoples have been defined and grouped according to their relationship with the five language families, and are found throughout the country [19]. Like the other Latin American countries, Venezuela has undergone an intense process of admixture, and in the 2011 census, 724,592 individuals self-declared as native Americans. This population is distributed among eight indigenous communities, the Amazonas, Anzoátegui, Apure, Bolívar, Delta Amacuro, Monagas, Sucre, and Zulia [20].

In the 2007 census of Peru, a total of 1786 indigenous communities were identified and mapped in 11 of the country's departments, with data being obtained on the populations and housing, with a total indigenous population of more than 4 million individuals [21]. In the United States, the Flathead Indian Reservation is home to three tribes, the Bitterroot Salish, the Upper Pend d'Oreille, and the Kootenai. The territories of these three tribes once covered all of western Montana and extended into parts of Idaho, British Columbia, and Wyoming [22]. Additional information on the ethnic groups and the countries in which they were studied is available in the Table S1. And Table 1 shows the published studies of the *CYP2D6* gene in Amerindian populations.

Table 1. The published studies of the *CYP2D6* gene in Amerindian populations compiled in the present study.

Variants	Population, Tribe, or Affiliation	Country	Reference
*1 *3 *4 *6, *7, *8	Tarahumaras, Purépechas, Tojolabales, Tzotziles, and Tzeltales	Mexico	[23]
*1, *2, *4, *5, *6, *10, *3, *17, *29, *35, *41	Mexicaneros, Seris, Guarijíos, Tepehuanos, Mayos, Huicholes, Tarahumaras, and Coras	Mexico	[24]
*2, *3, *4, *6, *10, *17, *29, *35, *41	Tarahumara, Tepehuana, Mexicanera, Huichol, Cora, Seri, Mayo, and Guarijía	Mexico	[25]
*3, *4, *6, *10	Tepehuano	Mexico	[26]
*2, *3, *4, *5, *6, *10, *17, *35, *41	Lacandones	Mexico	[27]
*5, *2A, *35, *41	Tapehuanos	Mexico	[28]
*2, *3, *4, *6, *10, *17, *35, *41	Tzotzil and Tzeltal	Mexico	[29]
*2, *3, *4, *5, *6, *10, *17, *29, *35, *41	Bri bri, Cabecar, Chorotega, Guatuso, Guaymi and Huetar; Chol, Huichol, Lacandon, Mayo, Mexicanero, Seri, Tarahumara, Tepehuano, Tzeltal, Tzoltil, Yaqui and Zoque; Ashaninka, Aymara, and Shima	Costa Rica, Mexico and Peru	[30]
*1, *2, *3, *4, *5, *6, *10, *17, *29, *35, *41	Bribri, Guaymi, Cabecar, Guatuso, Chorotega, and Huetar	Costa Rica	[31]
*1, *2, *3, *4, *5, *9, *10, *28, *33, *35, *41	Salish and Kootenai	United States of America	[32]
*2, *3, *4, *5, *6, *10	Bari, Panare, Pemon, Warao, and Wayuu	Venezuela	[33]
*1, *2, *3, *4, *5, *6, *8, *10, *12, *14, *15	Jujuy province, Wichi, Chorote, Toba, Mapuche, Tehuelche, Ayoreo, and Lengua	Argentina and Paraguay	[34]
*1, *2, *3, *4, *5, *9, *10	Mapuches	Chile	[35]

Many studies of American populations have demonstrated a degree of genetic heterogeneity in comparison with other ancestral populations, such as the Europeans and Africans, which also contributed to the formation of present-day New World populations [12,36,37]. The findings of these studies may contribute to the understanding of the interethnic variability of genetic polymorphisms in other populations, and their varying responses to drugs of clinical and therapeutic importance.

In recent years, there has been considerable interest in research on the diversity of drug-metabolizing enzyme genotypes and phenotypes, in particular *CYP2D6*, in different populations, especially those of European origin [38,39]. Other population groups have also been the target of this research, albeit on a much smaller scale. In particular, relatively few studies have focused on Amerindian populations, to investigate the role of gene variants in the absorption, distribution, metabolism, and excretion of drugs.

1.2. Objective

The present review aimed to compile the *CYP2D6* gene metabolization profiles of Amerindian populations through an extensive search of the available literature on molecular epidemiology, and to compare these data with those on other ethnic groups around the world.

2. Methods

2.1. Study Design

A literature search was used to identify studies on three research platforms (Google Scholar, Science Direct, and Pubmed), considering the 20-year period between 1998 and 2018. The search was based on two key terms, "CYP2D6 Amerindians" and "CYP2D6 native Americans", in all fields. The eligibility criteria were original papers in English, focusing on Amerindian populations, published between 1998 and 2018 (Figure 1).

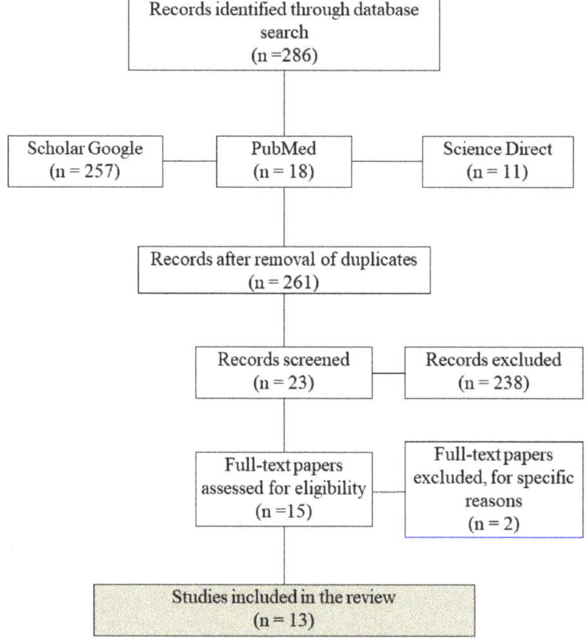

Figure 1. Flow diagram showing the number of records identified, included, and excluded from the present review.

2.2. Search Strategy

The following data were extracted from the selected papers: (i) the study population, (ii) the number of individuals analyzed, (iii) the polymorphisms of the CYP2D6 gene evaluated, and (iv) the allele and genotype frequencies. In all the papers selected here, the Amerindian populations were identified by self-declaration. The 1000 genomes platform was used as the reference database for the comparison of the allele frequencies of the most important gene mutations with those of other populations around the world.

3. Results

A total of 13 studies were evaluated and included in the present review. All the papers addressed the frequencies of the different *CYP2D6* alleles in Amerindian populations. Seven of the papers focused specifically on Amerindian populations from Mexico, while the others included populations from Argentina, Chile, Costa Rica, Mexico, Paraguay, Peru, and the United States (Table 1). The Native American populations, their country of origin, and the *CYP2D6* alleles recorded in each study are identified in Table 1. The populations were grouped by country to facilitate the discussion of the observed patterns.

Some of the papers identified in the literature search did not present data on the genotyping of the metabolization profiles, but only their classification in the Active Score (AS) system. In the AS approach, each allele is assigned a value of 0 (non-functioning), 0.5 (decreased), or 1 (normal). For alleles with two or more gene copies, the value of the allele is multiplied by the number of copies (e.g., the duplication *CYP2D6**1 × 2 is assigned an AS value of 2). The sum of the values assigned to the two alleles gives the AS score of the genotype. In this classification, an AS score of 0 was considered to represent a poor metabolizer (PM), while AS scores of 0.5–1 were classified as intermediate metabolizers (IM), scores of 1.5–2 as extensive metabolizers (EM), and scores of over 2 as ultrarapid metabolizers (UM) [40].

3.1. Synopsis of the Findings on the CYP2D6 Metabolization Profiles

3.1.1. Poor Metabolizers

Given the large number of polymorphisms found in the *CYP2D6* gene, the genotypes were characterized as the result of the interaction between haplotypes, with four principal metabolization phenotypes (Table 2). As the different variants have alternative functional consequences, individuals carrying these variants will have different levels of enzymatic activity. The metabolization profiles are classified according to the combination of alleles. The poor metabolizer (PM) profile is the result of the combination of two alleles that have a complete loss of function (no enzymatic activity), that is, null alleles due to mutations or the deletion of the gene [8,41]. The highest frequency of PM (30%) was recorded in Costa Rica [30,31], followed by Argentina/Paraguay, with 13% [34], and Venezuela [33] and the US [32], each with 6% (Figure 2).

Table 2. The different *CYP2D6* alleles recorded in the present review, and their respective variants and types of variation.

Allele (Arranged by Functional Consequence) Normal Activity	Variants	Variation Type
*1	None	-
*2	rs16947, rs1135840	Missense (R296C, S486T)
*33	rs28371717	Missense (A237S)
Increased Activity		
*1xN	Amplification of *1	-
*2xN	Amplification of *2	-
*53	rs1135822, rs1135823	Missense (F120I, A122S)

Table 2. *Cont.*

Allele (Arranged by Functional Consequence) Normal Activity	Variants	Variation Type
Decreased Activity		
*9	rs5030656	In-frame Deletion (K281del)
*10	rs1065852, rs1135840	Missense (P34S, S486T)
*17	rs16947, rs28371706	Missense (R296C, T107I)
*29	rs16947, rs1135840, rs61736512 and rs59421388	Missense (R296C, S486T, V136I and V338M)
*41	rs28371725	Splicing Defect
Inactivated		
*3	rs35742686	Frameshift
*4	rs3892097	Splicing Defect
*5	*CYP2D6* deletion	-
*6	rs5030655	Frameshift
*7	rs5030867	Missense (H324P)
*8	rs5030865	Stop gain (G169X)
*11	rs201377835	Splicing defect
*12	rs5030862	Missense (G42R)
*14	rs5030865	Missense (G169R)
*42	rs72549346	Frameshift
*62	rs730882171	Missense (R441C)

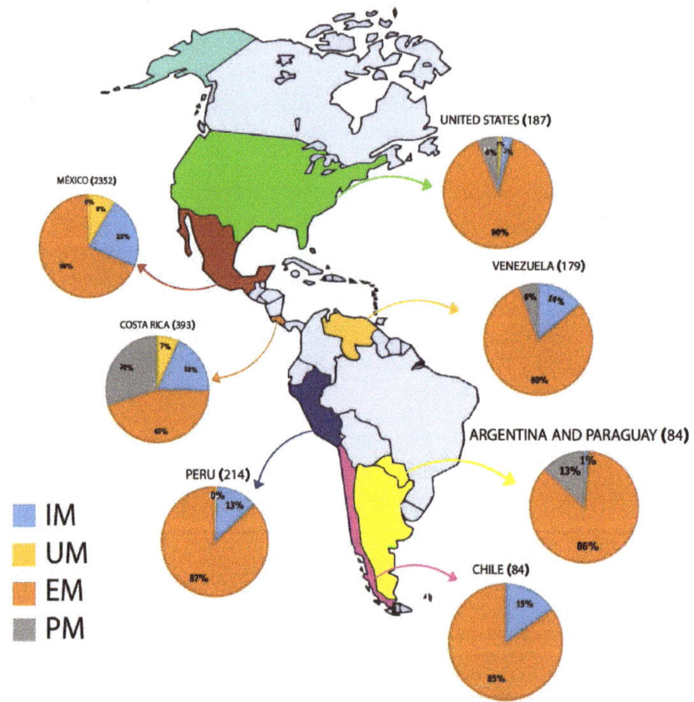

Figure 2. *CYP2D6* gene metabolization profile in Amerindian populations grouped by countries of America. IM: intermediate metabolizer; UM: ultrarapid metabolizer; EM: extensive metabolizer; PM: poor metabolizer.

The PM profile typically results in low levels of active metabolites of some medications, such as opioid analgesics, resulting in the reduced effectiveness of pain relief [42,43]. This indicates that individuals with the PM profile may require a modified therapeutic regimen or a follow-up for the diagnosis of the symptoms of insufficient pain relief [44].

3.1.2. Intermediate Metabolizers

Individuals with the IM phenotype have a combination of two alleles with decreased enzyme function or the presence of one non-functional allele together with a second allele with decreased function [8,41]. The IM profile is the second most frequent (19%) in the Amerindian populations included in the present review. The highest frequency of the IM profile was recorded in Mexico (22%) [23–30], followed by Costa Rica (18%) [30,31], Chile (15%) [35], Venezuela (14%) [33], Peru (13%) [30], the US (3%) [32], and Argentina/Paraguay (1%) [34] (see Figure 2).

The therapeutic implications of this profile vary among different drugs. In the case of opioid analgesics, such as codeine, individuals with the IM profile form reduced quantities of morphine from the medication, requiring a follow-up during treatment, although no change in drugs is required [31]. Despite the effects of the alleles with reduced enzymatic function, few data are available on the clinical impacts on drug treatment, response to therapy, or side effects.

3.1.3. Extensive Metabolizers

The EM profile does entail any significant alteration in enzymatic activity. This profile may be determined, for example, by two alleles with normal enzyme function or the combination of one normal allele and one with decreased function [8,41].

The EM profile was the most common in all countries (Figure 2), and the Amerindian populations from different countries present relatively homogeneous frequencies of this profile. The highest frequency (90%) was recorded in the populations from the US [32], followed by Peru (87%) [30], Argentina/Paraguay (86%) [34], Chile (85%) [35], and Venezuela (80%) [33]. However, much lower frequencies were recorded in Mexico (69%) and Costa Rica (45%) [24–31].

It is important to note here that, while the EM phenotype typically has normal enzymatic activity, tests are still required for the more accurate prediction of the catalytic activity of this metabolization profile [45,46].

3.1.4. Ultrarapid Metabolizers

Individuals with the UM profile carry at least one allele with increased enzyme function, in addition to an allele with normal function [41,47,48].

This profile was recorded in the Amerindian populations from three countries, Mexico, with a frequency of 9%, Costa Rica (7%), and the US, with 1% [24–31]. The highest frequency of this type of metabolizer has been recorded in an Ethiopian study (28.7%) [49]. Frequencies similar to those found in the Amerindian populations of Mexico have been recorded in Middle Eastern populations, i.e., 10.5% [7].

Despite the relatively low frequency of this profile in the study populations, its effects vary considerably, and may even be fatal. There have been a number of case reports of potentially fatal effects of standard doses of codeine in patients with the UM phenotype [50,51], which reinforces the need for the revision and alteration of the drugs used to treat these patients [52].

3.1.5. Allele Frequencies

The frequencies of the principal alleles that alter the functionality of the CYP2D6 enzyme in the Amerindian populations analyzed in the studies reviewed here were compared with those of the same alleles in other human populations from different regions of the world (data obtained from the 1000genomes database). As the frequencies of the amplifications and deletions are not provided in the 1000genomes database, they are now reviewed here. These plots are divided according to the functional implications of the alleles (normal, reduced, or loss of function) for comparison with populations from Europe, Africa, East and South Asia, and the Americas (Figure 3).

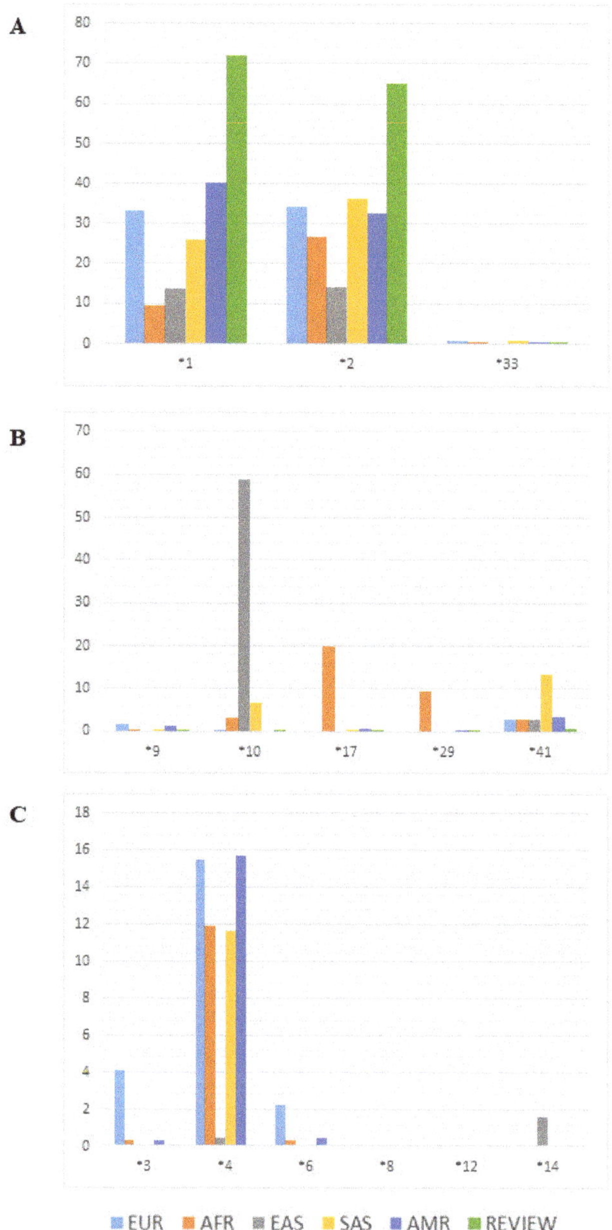

Figure 3. Allelic frequency of major variants of *CYP2D6* in world populations and in the Amerindian populations (data from review). (**A**) Variants with normal functional consequence; (**B**) variants with decreased functional consequence; (**C**) variants with inactivated functional consequence. EUR: European populations; AFR: African populations; EAS: East Asian populations; SAS: South Asian populations; AMR: American mixed populations; REVIEW: Amerindian populations from the review.

The frequencies of the alleles with normal function are shown in Figure 3A. In general, the Amerindian populations have a high frequency of normal alleles, in particular *1 and *2, in comparison with the other populations. Admixed populations with a major Native American ancestry component have frequencies similar to those of the Amerindian populations presented here. Worldwide, 77–92% of individuals have at least one copy of a normal allele (*1 or *2) or two partially functional alleles [45].

The Amerindian populations presented frequencies close to 0% for all variants of the alleles with reduced function (Figure 3B). The *CYP2D6*10* allele occurs at high frequencies in South Asian populations, while *CYP2D6*41*, which is caused by a splicing defect, has a frequency of less than 10%. The *CYP2D6*17* allele, which results from two missense mutations, has a frequency of 20% in the African populations, while *CYP2D6*29*, which has four missense mutations, has a frequency of 10% in these populations.

The *CYP2D6*4* variant, which is caused by a splicing defect, which inactivates the product of the *CYP2D6* gene, has frequencies ranging from 11.6% to 15.7% in most populations, except in East Asians and in the Amerindian populations reviewed here, in which the allele frequency was less than 1%. The frequencies of the other variants that cause the inactivation of the gene product were generally low, and did not vary greatly among populations (Figure 3C). However, *CYP2D6*3* had a frequency of 4% in the European populations, while *CYP2D6*14* had a frequency of 2%. The East Asian populations also presented frequencies of over 1% for the *CYP2D6*6* variant.

4. Discussion

The *CYP2D6* gene has many polymorphisms, resulting in a high level of inter-individual variability [8]. The molecular *CYP2D6* profile may alter the therapeutic efficacy of different drugs, which has led a number of international drug regulatory agencies to recommend the use of *CYP2D6* polymorphisms as biomarkers for the design of therapies based on antidepressants, antipsychotics, antiarrhythmics, opioid analgesics, anticancer agents, and other classes of drug [45,53]. A large number of the variants of the *CYP2D6* gene have important implications for the application of clinical therapy, and depending on their genomic profile, a patient may have a high risk of adverse reactions or even a failure of the therapy. The molecular profile of the *CYP2D6* gene varies considerably among populations [7].

The principal ancestral human populations—Europeans, Asians, Amerindians, and Africans—present considerable genetic diversity, which implies major fluctuations in the frequencies of important pharmacogenetic polymorphisms [54]. The traditional populations of the Americas have a long history of geographical isolation, and a differentiated genetic makeup in comparison with populations from other regions of the world [24]. The common origin of Native Americans and their autochthonous biological and cultural evolution combine to make the New World an excellent model for studies of the co-evolution between genes and cultures [54]. The papers selected for the present review permitted the compilation of the *CYP2D6* gene metabolization profiles of the Amerindian populations of a number of different countries in the Americas. The four metabolizing profiles (PM, IM, EM, and UM) had varying frequencies in the different countries (Figure 2).

The EM and IM metabolization profiles were the most frequent in all countries surveyed in the present review. Some drug regulatory agencies recommend the follow-up for patients using drugs metabolized by *CYP2D6* that have an IM profile. In the specific case of codeine, for example, there is evidence that some adverse effects do not vary significantly between poor and extensive metabolizers, which re-emphasizes the need for testing to determine the catalytic activity of individuals with these profiles to guarantee the best possible therapeutic strategy [42,45]. One important question here is that the extensive and intermediate metabolization profiles can be "converted" into a slow metabolization profile when exposed to other xenobiotics, such as alkaloid drugs or herbal medicines, which are known to be potent *CYP2D6* inhibitors, and are commonly used by Amerindian populations [48].

This is important given the high frequencies of the EM and IM profiles in these traditional populations and their extensive use of xenobiotics.

The ultrarapid metabolization (UM) profile has a relatively low frequency in the general Amerindian population (7%), although it is slightly higher in Mexican populations (9%) than in the other countries [24–30], probably due to the isolation of Mexican mestizos [24,30]. The UM profile was also substantial (7%) in Costa Rica [30,31]. These frequencies are similar to those described in other populations, such as that of Spain (6.1%) [50]. Individuals with this phenotype metabolize drugs faster than normal, which means that drugs taken at the standard dose may not have the intended therapeutic effects. Worse still, these individuals may also develop adverse reactions due to the formation of relatively large quantities of metabolites, between 10 and 30 times larger than normal [55,56].

The PM profile was relatively infrequent in the general Amerindian population (4%), a frequency similar to that recorded in Europe (6.52%) by Llerena et al. (2014) [7]. Low frequencies of slow metabolizers were recorded in a number of Native Mexican populations [23,24,26,27]. Sosa-Macías et al. (2010) found that Mexican Amerindian populations, in particular those of the Taphuanos community, had a more homogeneous slow metabolism profile distribution in comparison with Mexican mestizos [28]. This may be accounted for by the Tapehuano group and its low levels of miscegenation with Mexican mestizos. Individuals with the slow metabolizing phenotype have a potential risk of accumulating toxic metabolites when undergoing treatment with opioid analgesics, such as morphine and tramadol, and during long term therapy with antipsychotics and antihypertensive agents. In addition, they are prone to a lack of effective pharmaceutical action in the case of pro-drugs metabolized by the CYP2D6 enzyme, such as codeine and tamoxifen [29]. Given this, individuals of the Mexican Amerindian ethnic group have a lower risk of developing adverse drug reactions or therapeutic failure when treated with the aforementioned drugs.

The mean frequency of the alleles associated with the PM profile in Costa Rica, Argentina, Paraguay, the US, and Venezuela was 13.75%, with some variation being found among the Amerindian populations of these countries [30–34]. Naranjo et al. (2018) found that 10.2% of the indigenous Costa Rican population were slow metabolizers [30], a similar percentage to that recorded by Céspedes-Garro et al. (2014) for the same ethnic group [31]. In a survey of Venezuelan Amerindian populations, Grimam et al. (2012) found slow metabolizers in only one of the five populations analyzed [33]. Based on data from Argentinian and Paraguayan populations, Bailliet et al. (2007) concluded that some of the mutations that make up the PM genotype are founder variants brought to America by the early Asian settlers [34]. The different alleles of the slow metabolism profiles are described as important therapy predictors by Fohner et al. (2013), who evaluated native populations from the US, and suggested that CYP2D6 activity may be decreased (with high frequencies of the non-functional *CYP2D6*4* and *CYP2D6*41* variants) in 9.09% of the patients from the Salish and Kootenai tribes [32].

The results of the present review indicate that the frequencies of some of the alleles present in the Amerindian populations are similar to those of populations from other regions of the world (Figure 3). The majority of the *CYP2D6* alleles are shared by most of the world's populations. Even so, a number of different evolutionary factors may have contributed to the establishment of geographical gradients in the distribution of some alleles, which occur at high frequencies in certain, specific regions of the world [57]. During a period of population expansion, in particular the initial wave, some rare alleles or haplotypes may become relatively common through founder effects [58].

A number of factors may be associated with the variation in allele frequencies of drug-metabolizing enzymes, such as CYP2D6, and the resulting metabolization profiles of Amerindian populations. Nebert (1997) identifies two possible selective pressures that may have determined this variation—differences in diet, which evolved over thousands of years, and the evolution of balanced polymorphisms, including alleles that confer resistance to bacterial or viral infections [59]. Considerable research efforts have been invested in the identification of predictive markers of therapeutic conduct for the development of personalized treatment protocols. In general, Amerindian populations have a different genetic profile than those of other populations around the world, which reinforces the need for specific studies of these ethnic groups for the identification of novel biomarkers relevant to the standardization of therapeutic strategies. The present review compiled the major publications available on the *CYP2D6* gene in Native American populations. As there is a clear under-representation of this ethnic group in pharmacogenomic studies, it will be essential to determine the *CYP2D6* profiles of a much larger number of Native American populations to support the development of systematic public health strategies.

It is important to note that substrate specificity and drug dosage are only two of a range of factors that contribute to the response of an individual to a given drug. Research groups, such as the Ibero-American Network of Pharmacogenetics and Pharmacogenomics (RIBEF) consortium, were established for the study of pharmacologically important genes in Latin American populations. The present review represents a pioneering compilation of the results of the available papers on the variation in the *CYP2D6* gene in Amerindian populations, and highlights the importance of this research for the development of guidelines for the management of therapeutic strategies in these populations. The major deficiency of the available papers is the lack of data from some countries, most notably, Brazil, Canada, and Colombia. These considerations reinforce the need for further, more extensive studies on the pharmacogenetics of Amerindian populations.

Supplementary Materials: The owing are available online at http://www.mdpi.com/2073-4425/11/3/262/s1, Table S1: Population number and geographic location of the ethnic groups described in the review.

Author Contributions: L.P.C.L. designed the study, processed the data and wrote the review. T.P.S. developed the review methods and produced the figures. J.C.G.R. and M.R.F. contributed to the development of the methods. S.S. and N.P.C.S. were the coordinators of the project. All authors have read and agreed to the published version of the manuscript.

Funding: This research was funded by CNPq (Conselho Nacional de Desenvolvimento Científico e Tecnológico), FAPESPA (Fundação Amazônica de Amparo a Estudos e Pesquisas), and CAPES (Coordenação de Aperfeiçoamento de Pessoal de Nível Superior). The APC was funded by UFPA (Universidade Federal do Pará). These funding agencies played no role in the study design, data collection and analysis, or the decision to publish, or the preparation of the manuscript.

Acknowledgments: We acknowledge the Research Oncology Center and the Oncology and Medical Sciences PostGraduate Program for the support of the research.

Conflicts of Interest: The authors declare that they have no competing interests.

References

1. Wendt, F.R.; Sajantila, A.; Moura-Neto, R.S.; Woerner, A.E.; Budowle, B. Full-gene haplotypes refine CYP2D6 metabolizer phenotype inferences. *Int. J. Leg. Med.* **2017**, *132*, 1007–1024. [CrossRef]

2. Ingelman-Sundberg, M. Genetic polymorphisms of cytochrome P450 2D6 (CYP2D6): Clinical consequences, evolutionary aspects and functional diversity. *Pharm. J.* **2005**, *5*, 6–13. [CrossRef]

3. Zanger, U.M.; Schwab, M. Cytochrome P450 enzymes in drug metabolism: Regulation of gene expression, enzyme activities, and impact of genetic variation. *Pharmacol. Ther.* **2013**, *138*, 103–141. [CrossRef]

4. Kimura, S.; Umeno, M.; Skoda, R.C.; Meyer, U.A.; Gonzalez, F.J. The human debrisoquine 4-hydroxylase (CYP2D) locus: Sequence and identification of the polymorphic CYP2D6 gene, a related gene, and a pseudogene. *Am. J. Hum. Genet.* **1989**, *45*, 889–904.

5. Henderson, L.; Claw, K.; Woodahl, E.L.; Robinson, R.F.; Boyer, B.; Burke, W.; Thummel, K. P450 Pharmacogenetics in Indigenous North American Populations. *J. Pers. Med.* **2018**, *8*, 9. [CrossRef]

6. He, Z.-X.; Chen, X.-W.; Zhou, Z.-W.; Zhou, S.-F. Impact of physiological, pathological and environmental factors on the expression and activity of human cytochrome P450 2D6 and implications in precision medicine. *Drug Metab. Rev.* **2015**, *47*, 470–519. [CrossRef]

7. Llerena, A.; Naranjo, M.E.G.; Rodrigues-Soares, F.; Peñas-Lledó, E.E.; Fariñas, H.; Tarazona-Santos, E. Interethnic variability of CYP2D6 alleles and of predicted and measured metabolic phenotypes across world populations. *Expert Opin. Drug Metab. Toxicol.* **2014**, *10*, 1569–1583. [CrossRef]

8. Zhou, Y.; Ingelman-Sundberg, M.; Lauschke, V.M. Worldwide Distribution of Cytochrome P450 Alleles: A Meta-analysis of Population-scale Sequencing Projects. *Clin. Pharmacol. Ther.* **2017**, *102*, 688–700. [CrossRef] [PubMed]

9. McGraw, J.E.; Waller, D. Cytochrome P450 variations in different ethnic populations. *Expert Opin. Drug Metab. Toxicol.* **2012**, *8*, 371–382. [CrossRef] [PubMed]

10. Sistonen, J.; Fuselli, S.; Palo, J.U.; Chauhan, N.; Padh, H.; Sajantila, A. Pharmacogenetic variation at CYP2C9, CYP2C19, and CYP2D6 at global and microgeographic scales. *Pharm. Genom.* **2009**, *19*, 170–179. [CrossRef] [PubMed]

11. U.S. Census Bureau. *2010 Census of Population and Housing, Population and Housing Unit Counts, CPH-2-1*; United States Summary U.S. Government Printing Office: Washington, DC, USA, 2012.

12. Salzano, F.M.; Sans, M. Interethnic admixture and the evolution of Latin American populations. *Genet. Mol. Boil.* **2014**, *37*, 151–170. [CrossRef] [PubMed]

13. Salzano, F.M.; Bortolini, M.C. *Evolution and Genetics of Latin American Populations*; Cambridge University Press: Cambridge, UK, 2002.

14. Boege, E. *Regiones, Territorio, Lenguas y Cultura de los Pueblos Indígenas; El Patrimonio Biocultural de los Pueblos Indígenas de México. Hacia la Conservación in situ de la Biodiversidad y Agrodiversidad en los Territorios Indígenas, Eckart Boege*; Instituto Nacional de Antropología e Historia-Comisión Nacional para el Desarrollo de los Pueblos Indígenas: Ciudad de México, México, 2008; pp. 49–63.

15. Ministerio de Desarrollo Social. Ministerio de Desarrollo Social. Casen 2015. Pueblos Indígenas. Síntesis de Resultados. Available online: http://observatorio.ministeriodesarrollosocial.gob.cl/casen-multidimensional/casen/docs/CASEN_2015_Resultados_pueblos_indigenas.pdf (accessed on 10 April 2018).

16. Arnaiz-Villena, A.; Juarez, I.; Lopez-Nares, A.; Palacio-Grüber, J.; Vaquero, C.; Callado, A.; H-Sevilla, A.; Rey, D.; Martin-Villa, J.M.; Juarez, I.; et al. Frequencies and significance of HLA genes in Amerindians from Chile Cañete Mapuche. *Hum. Immunol.* **2019**, *80*, 419–420. [CrossRef] [PubMed]

17. Azofeifa, J.; Ruiz-Narváez, E.A.; Leal, A.; Gerlovin, H.; Rosero-Bixby, L. Amerindian ancestry and extended longevity in Nicoya, Costa Rica. *Am. J. Hum. Biol.* **2018**, *30*, e23055. [CrossRef] [PubMed]

18. Instituto Nacional de Estadística y Censos-Costa Rica. *X Censo Nacional de Población y VI de Vivienda 2011: Características Sociales y Demográficas Tomo II/Instituto Nacional de Estadística y Censos*, 1st ed.; INEC: San José, CA, USA, 2012; 340p.

19. Dirección General de Estadística, Encuestas y Censos (DGEEC)–III Censo Nacional de Población Y Viviendas para Pueblos Indígenas. 2012. Available online: https://www.dgeec.gov.py/Publicaciones/Biblioteca/censo%20indigena%202012/Presentacion%20resultados%2019%2007%2013.pdf (accessed on 27 January 2020).

20. Instituto Nacional de Estadística. *La Población Indígena de Venezuela. Censo 2011*; Instituto Nacional de Estadística: Caracas, Venezuela, 2013; Volume 1, 15p.

21. Instituto Nacional de Estadística e Informática. *Censos Nacionales 2007: XI de Población y vi de Vivienda-Resumen Ejecutivo Resultados Definitivos de las Comunidades Indígenas*; Dirección Nacional de Censos y Encuestas: Lima, Peru, 2009; 168p.

22. Bigart, R.; Woodcock, C. (Eds.) *the Name of the Salish and Kootenai Nation: The 1855 Hell Gate Treaty and the Origin of the Flathead Indian Reservation*; Salish Kootenai College Press: Pablo, MT, USA, 1996.

23. Salazar-Flores, J.; Torres-Reyes, L.A.; Martínez-Cortés, G.; Castellanos, R.R.; Sosa-Macías, M.; Muñoz-Valle, J.F.; González-González, C.; Ramirez, A.; Roman, R.; Mendez, J.L.; et al. Distribution of CYP2D6 and CYP2C19 Polymorphisms Associated with Poor Metabolizer Phenotype in Five Amerindian Groups and Western Mestizos from Mexico. *Genet. Test. Mol. Biomarkers* **2012**, *16*, 1098–1104. [CrossRef] [PubMed]

24. Lazalde-Ramos, B.P.; Martinez-Fierro, M.L.; Galaviz-Hernandez, C.; Garza-Veloz, I.; Naranjo, M.E.G.; Sosa-Macías, M.; Llerena, A.; Llerena, A. CYP2D6 gene polymorphisms and predicted phenotypes in eight indigenous groups from northwestern Mexico. *Pharmacogenomics* **2014**, *15*, 339–348. [CrossRef]

25. De Andrés, F.; Macías, M.S.; Ramos, B.P.L.; Naranjo, M.-E.G.; Llerena, A. CYP450 Genotype/Phenotype Concordance in Mexican Amerindian Indigenous Populations–Where to from Here for Global Precision Medicine? *OMICS J. Integr. Boil.* **2017**, *21*, 509–519. [CrossRef]

26. Sosa-Macías, M.; Elizondo, G.; Flores-Pérez, C.; Flores-Pérez, J.; Bradley-Alvarez, F.; Alanis-Bañuelos, R.E.; Lares-Asseff, I. CYP2D6Genotype and Phenotype in Amerindians of Tepehuano Origin and Mestizos of Durango, Mexico. *J. Clin. Pharmacol.* **2006**, *46*, 527–536. [CrossRef]

27. López-López, M.; Peñas-Lledó, E.; Dorado, P.; Ortega, A.; Corona, T.; Ochoa, A.; Yescas, P.; Alonso, E.; Llerena, A.; Llerena, A. CYP2D6 genetic polymorphisms in Southern Mexican Mayan Lacandones and Mestizos from Chiapas. *Pharmacogenomics* **2014**, *15*, 1859–1865. [CrossRef]

28. Sosa-Macías, M.; Dorado, P.; Alanis-Bañuelos, R.E.; Llerena, A.; Lares-Asseff, I. Influence of CYP2D6 deletion, multiplication, −1584C→ G, 31G→ A and 2988G→ a gene polymorphisms on dextromethorphan metabolism among Mexican tepehuanos and mestizos. *Pharmacology* **2010**, *86*, 30–36. [CrossRef]

29. Perez-Paramo, Y.X.; Hernandez-Cabrera, F.; Dorado, P.; Llerena, A.; Muñoz-Jimenez, S.; Ortiz-Lopez, R.; Rojas-Martinez, A.; Llerena, A. Interethnic relationships of CYP2D6 variants in native and Mestizo populations sharing the same ecosystem. *Pharmacogenomics* **2015**, *16*, 703–712. [CrossRef]

30. Naranjo, M.-E.G.; Rodrigues-Soares, F.; Peñas-Lledó, E.E.; Tarazona-Santos, E.; Fariñas, H.; Rodeiro, I.; Teran, E.; Grazina, M.; Moya, G.E.; López-López, M.; et al. Interethnic Variability in CYP2D6, CYP2C9, and CYP2C19 Genes and Predicted Drug Metabolism Phenotypes among 6060 Ibero- and Native Americans: RIBEF-CEIBA Consortium Report on Population Pharmacogenomics. *OMICS J. Integr. Boil.* **2018**, *22*. [CrossRef] [PubMed]

31. Céspedes-Garro, C.; Jiménez-Arce, G.; Naranjo, M.-E.G.; Barrantes, R.; Llerena, A. Ethnic background and CYP2D6 genetic polymorphisms in Costa Ricans. *Rev. Biol. Trop.* **2014**, *62*, 1659. [CrossRef] [PubMed]

32. Fohner, A.; Muzquiz, L.I.; Austin, M.A.; Gaedigk, A.; Gordon, A.; Thornton, T.; Rieder, M.J.; Pershouse, M.A.; Putnam, E.A.; Howlett, K.; et al. Pharmacogenetics in American Indian populations: Analysis of CYP2D6, CYP3A4, CYP3A5, and CYP2C9 in the Confederated Salish and Kootenai Tribes. *Pharm. Genom.* **2013**, *23*, 403. [CrossRef] [PubMed]

33. Grimán, P.; Moran, Y.; Valero, G.; Loreto, M.; Borjas, L.; Chiurillo, M.A. CYP2D6 gene variants in urban/admixed and Amerindian populations of Venezuela: Pharmacogenetics and anthropological implications. *Ann. Hum. Boil.* **2012**, *39*, 137–142. [CrossRef] [PubMed]

34. Bailliet, G.; Santos, M.; Alfaro, E.; Dipierri, J.E.; Demarchi, D.; Carnese, F.; Bianchi, N. Allele and genotype frequencies of metabolic genes in Native Americans from Argentina and Paraguay. *Mutat. Res. Toxicol. Environ. Mutagen.* **2007**, *627*, 171–177. [CrossRef]

35. Muñoz, S.; Vollrath, V.; Vallejos, M.P.; Miquel, J.F.; Covarrubias, C.; Raddatz, A.; Chianale, J. Genetic polymorphisms of CYP2D6, CYP1A1 and CYP2E1 in the South-Amerindian population of Chile. *Pharmacogenetics* **1998**, *8*, 343–351. [PubMed]

36. Pena, S.D.J.; Di Pietro, G.; Fuchshuber-Moraes, M.; Genro, J.P.; Hutz, M.H.; Kehdy, F.D.S.G.; Kohlrausch, F.; Magno, L.A.V.; Montenegro, R.C.; Moraes, M.O.; et al. The Genomic Ancestry of Individuals from Different Geographical Regions of Brazil Is More Uniform Than Expected. *PLoS ONE* **2011**, *6*, e17063. [CrossRef]

37. Wang, S.; Ray, N.; Rojas, W.; Parra, M.V.; Bedoya, G.; Gallo, C.; Poletti, G.; Mazzotti, G.; Hill, K.; Hurtado, A.M.; et al. Geographic Patterns of Genome Admixture in Latin American Mestizos. *PLoS Genet.* **2008**, *4*, e1000037. [CrossRef]

38. Céspedes-Garro, C.; Fricke-Galindo, I.; Naranjo, M.E.G.; Rodrigues-Soares, F.; Fariñas, H.; De Andrés, F.; López-López, M.; Peñas-Lledó, E.M.; Llerena, A.; Llerena, A. Worldwide interethnic variability and geographical distribution of CYP2C9 genotypes and phenotypes. *Expert Opin. Drug Metab. Toxicol.* **2015**, *11*, 1893–1905. [CrossRef]

39. Fricke-Galindo, I.; Céspedes-Garro, C.; Rodrigues-Soares, F.; Naranjo, M.E.G.; Delgado, Á.; de Andrés, F.; López-López, M.; Peñas-Lledó, E.; LLerena, A. Interethnic variation of CYP2C19 alleles, "predicted" phenotypes and "measured" metabolic phenotypes across world populations. *Pharm. J.* **2016**, *16*, 113–123. [CrossRef]

40. Gaedigk, A.; Dinh, J.C.; Jeong, H.; Prasad, B.; Leeder, S. Ten Years' Experience with the CYP2D6 Activity Score: A Perspective on Future Investigations to Improve Clinical Predictions for Precision Therapeutics. *J. Pers. Med.* **2018**, *8*, 15. [CrossRef]

41. Gaedigk, A.; Sangkuhl, K.; Whirl-Carrillo, M.; Klein, T.; Leeder, J.S. Prediction of CYP2D6 phenotype from genotype across world populations. *Genet. Med.* **2017**, *19*, 69–76. [CrossRef] [PubMed]

42. Eckhardt, K.; Li, S.; Ammon, S.; Schänzle, G.; Mikus, G.; Eichelbaum, M. Same incidence of adverse drug events after codeine administration irrespective of the genetically determined differences in morphine formation. *Pain* **1998**, *76*, 27–33. [CrossRef]

43. Slanar, O.; Dupal, P.; Matouskova, O.; Vondrackova, H.; Pafko, P.; Perlík, F. Tramadol efficacy in patients with postoperative pain in relation to CYP2D6 and MDR1 polymorphisms. *Bratisl. Lek. Listy* **2012**, *113*, 152–155. [CrossRef]

44. Swen, J.J.; Nijenhuis, M.; De Boer, A.; Grandia, L.; Der Zee, A.H.M.-V.; Mulder, H.; Rongen, G.A.P.J.M.; Van Schaik, R.H.N.; Schalekamp, T.; Touw, D.J.; et al. Pharmacogenetics: From Bench to Byte—An Update of Guidelines. *Clin. Pharmacol. Ther.* **2011**, *89*, 662–673. [CrossRef] [PubMed]

45. Crews, K.R.; Gaedigk, A.; Dunnenberger, H.; Leeder, J.S.; Klein, T.E.; Caudle, K.E.; Haidar, C.E.; Shen, D.D.; Callaghan, J.T.; Sadhasivam, S.; et al. Clinical Pharmacogenetics Implementation Consortium Guidelines for Cytochrome P450 2D6 Genotype and Codeine Therapy: 2014 Update. *Clin. Pharmacol. Ther.* **2014**, *95*, 376–382. [CrossRef]

46. Bertilsson, L.; Dahl, M.; Dalén, P.; Al-Shurbaji, A. Molecular genetics of CYP2D6: Clinical relevance with focus on psychotropic drugs. *Br. J. Clin. Pharmacol.* **2002**, *53*, 111–122. [CrossRef] [PubMed]

47. Zhou, S.-F. Polymorphism of Human Cytochrome P450 2D6 and Its Clinical Significance. *Clin. Pharm.* **2009**, *48*, 761–804. [CrossRef]

48. Yu, C.Y.; Ang, G.Y.; Subramaniam, V.; James, R.M.J.; Ahmad, A.; Rahman, T.A.; Nor, F.M.; Shaari, S.A.; Teh, L.K.; Salleh, H. Inference of the Genetic Polymorphisms of CYP2D6 in Six Subtribes of the Malaysian Orang Asli from Whole-Genome Sequencing Data. *Genet. Test. Mol. Biomarkers* **2017**, *21*, 409–415. [CrossRef]

49. Aklillu, E.; Persson, I.; Bertilsson, L.; Johansson, I.; Rodrigues, F.; Ingelman-Sundberg, M. Frequent distribution of ultrarapid metabolizers of debrisoquine in an ethiopian population carrying duplicated and multiduplicated functional CYP2D6 alleles. *J. Pharmacol. Exp. Ther.* **1996**, *278*, 441–446.

50. Gasche, Y.; Daali, Y.; Fathi, M.; Chiappe, A.; Cottini, S.; Dayer, P.; Desmeules, J. Codeine Intoxication Associated with Ultrarapid CYP2D6 Metabolism. *N. Engl. J. Med.* **2004**, *351*, 2827–2831. [CrossRef] [PubMed]

51. Ciszkowski, C.; Madadi, P.; Phillips, M.S.; Lauwers, A.E.; Koren, G. Codeine, Ultrarapid-Metabolism Genotype, and Postoperative Death. *N. Engl. J. Med.* **2009**, *361*, 827–828. [CrossRef] [PubMed]

52. Dean, L. Codeine therapy and CYP2D6 genotype. In *Medical Genetics Summaries (Internet)*; National Center for Biotechnology Information (US). Available online: https://www.ncbi.nlm.nih.gov/books/NBK100662/ (accessed on 15 December 2019).

53. Gardiner, S.J.; Begg, E.J. Pharmacogenetics, Drug-Metabolizing Enzymes, and Clinical Practice. *Pharmacol. Rev.* **2006**, *58*, 521–590. [CrossRef] [PubMed]

54. Jittikoon, J.; Mahasirimongkol, S.; Charoenyingwattana, A.; Chaikledkaew, U.; Tragulpiankit, P.; Mangmool, S.; Inunchot, W.; Somboonyosdes, C.; Wichukchinda, N.; Sawanpanyalert, P.; et al. Comparison of genetic variation in drug ADME-related genes in Thais with Caucasian, African and Asian HapMap populations. *J. Hum. Genet.* **2016**, *61*, 119–127. [CrossRef] [PubMed]

55. Peñas-Lledó, E.M.; Dorado, P.; Agüera, Z.; Gratacós, M.; Estivill, X.; Fernández-Aranda, F.; LLerena, A. CYP2D6 polymorphism in patients with eating disorders. *Pharm. J.* **2012**, *12*, 173–175. [CrossRef]

56. Dalén, P.; Frengell, C.; Dahl, M.-L.; Sjöqvist, F. Quick onset of severe abdominal pain after codeine in an ultrarapid metabolizer of debrisoquine. *Ther. Drug Monit.* **1997**, *19*, 543–544. [CrossRef]

57. Sistonen, J.; Sajantila, A.; Lao, O.; Corander, J.; Barbujani, G.; Fuselli, S. CYP2D6 worldwide genetic variation shows high frequency of altered activity variants and no continental structure. *Pharm. Genom.* **2007**, *17*, 93–101.

58. Excoffier, L.; Ray, N. Surfing during population expansions promotes genetic revolutions and structuration. *Trends Ecol. Evol.* **2008**, *23*, 347–351. [CrossRef]

59. Nebert, D.W. Polymorphisms in drug-metabolizing enzymes: What is their clinical relevance and why do they exist? *Am. J. Hum. Genet.* **1997**, *60*, 265–271.

 *genes*

Article

Rising Roles of Small Noncoding RNAs in Cotranscriptional Regulation: In Silico Study of miRNA and piRNA Regulatory Network in Humans

Massimiliano Chetta [1,*,†], **Lorena Di Pietro** [2,*,†], **Nenad Bukvic** [3,‡] and **Wanda Lattanzi** [2,4,‡]

1   U.O.C. Genetica Medica e di Laboratorio, Ospedale Antonio Cardarelli, 80131 Napoli, Italy
2   Dipartimento Scienze della Vita e Sanità Pubblica, Sezione di Biologia Applicata, Università Cattolica del Sacro Cuore, 00168 Rome, Italy; wanda.lattanzi@unicatt.it
3   UOC Lab. di Genetica Medica, Azienda Ospedaliero Universitaria Consorziale Policlinico di Bari, 70124 Bari, Italy; nenadbukvic@virgilio.it
4   Fondazione Policlinico Universitario A. Gemelli IRCCS, 00168 Rome, Italy
*   Correspondence: mchetta@unisa.it (M.C.); lorena.dipietro@unicatt.it (L.D.P.); Tel.: +39-333187551 (M.C.); +39-0630154464 (L.D.P.)
†   M.C. and L.D.P. contributed equally.
‡   W.L. and N.B. share senior authorship.

Received: 20 March 2020; Accepted: 27 April 2020; Published: 29 April 2020

Abstract: Gene expression regulation is achieved through an intricate network of molecular interactions, in which trans-acting transcription factors (TFs) and small noncoding RNAs (sncRNAs), including microRNAs (miRNAs) and PIWI-interacting RNAs (piRNAs), play a key role. Recent observations allowed postulating an interplay between TFs and sncRNAs, in that they may possibly share DNA-binding sites. The aim of this study was to analyze the complete subset of miRNA and piRNA sequences stored in the main databases in order to identify the occurrence of conserved motifs and subsequently predict a possible innovative interplay with TFs at a transcriptional level. To this aim, we adopted an original in silico workflow to search motifs and predict interactions within genome-scale regulatory networks. Our results allowed categorizing miRNA and piRNA motifs, with corresponding TFs sharing complementary DNA-binding motifs. The biological interpretation of the gene ontologies of the TFs permitted observing a selective enrichment in developmental pathways, allowing the distribution of miRNA motifs along a topological and chronological frame. In addition, piRNA motifs were categorized for the first time and revealed specific functional implications in somatic tissues. These data might pose experimental hypotheses to be tested in biological models, towards clarifying novel in gene regulatory routes.

Keywords: gene expression regulation; miRNA; piRNA; transcriptional modulation; binding motifs; regulatory pathways; epigenetics; in silico

---

1. Introduction

Roughly 90% of the genome is transcribed. Protein-coding (structural) genes account for as little as ~2% of the genome, with a much larger portion of the transcribed content being represented by noncoding RNAs (ncRNAs) [1,2]. Since the original characterization of the first ncRNA, a transfer RNA (tRNA) purified from yeast in 1965 [3], the conceptualization of the "RNA world" [4] and the discovery of different types of ncRNAs (including ribosomal RNAs, long and small noncoding RNAs, and circular RNAs) has improved significantly. The understanding of the roles and mechanisms of action of noncoding RNAs (ncRNAs) has been progressively increasing throughout this postgenomic era, completely revolutionizing the idea of "junk DNA" [1,2]. In particular, most small noncoding RNAs (sncRNAs) discovered so far (i.e., microRNAs (miRNAs), PIWI-interacting RNAs (piRNAs), tRNA-derived stress-induced small RNAs (tiRNAs), small-interfering RNAs (siRNAs)) are responsible for a variety of regulatory mechanisms affecting gene expression at multiple levels.

Gene expression is regulated in a tissue-specific fashion with time-related rhythms and stimuli-responsive mechanisms. This is achieved through the fine-tuning orchestrated by a number of molecular interactors and epigenetic regulators that act at different levels. These include known DNA trans-acting elements, such as transcription factors (TFs), mostly binding to promoter elements upstream of genes, and transcriptional activators and repressors, which interact with enhancer and silencer regions, respectively. In addition, different classes of sncRNAs, including miRNAs and piRNAs, perform a dynamic epigenetic regulation and surveillance of genomic integrity and functions, acting on the genome scale and determining a dense and complex regulatory network [5].

miRNAs are short, single-stranded noncoding RNAs (20–22 nucleotides in length) that are typically transcribed by the RNA polymerase II. miRNAs are encoded by the major regulatory gene family in eukaryotic cells and operate in multiple signaling pathways [6,7]. In particular, the control of mRNA stability and translation is the best known and characterized miRNA function. According to the consolidated model, mature miRNAs are found and act in the cytoplasm, where they recognize and bind complementary cis-regulatory elements usually located in the 3′UTR of the target mRNAs, through a specific short (6–7 nucleotides) sequence known as 'seed sequence' in their 5′ end. This interaction determines the degradation of the mRNA or the repression of its translation [5]. miRNAs are commonly categorized in family groups based on their sequence conservation, with specific regard to the seed sequence. Therefore, the miRNA family members considered to date tend to share similar targets and are usually involved in interconnected pathways [8].

Increasing evidence points towards the idea that mature miRNAs are also enriched in the nucleus, suggesting noncanonical roles, i.e., in transcriptional silencing or activation and alternative splicing [9–11]. The conditions through which miRNAs elicit their nuclear function have not been fully clarified, and the mechanisms that regulate miRNAs' translocation into the nucleus are also extensively debated [11]. It has been reported that miRNAs can induce gene expression by directly binding to the promoter region, as TFs, through sequence complementarity [9,12].

piRNAs are 24–32-nucleotide-long noncoding RNAs, named after their association with the argonaute subfamily PIWI proteins, originally identified in mutants affected by asymmetric division of stem cells in the Drosophila germline [13]. piRNAs are the largest subgroups of sncRNAs recognized in animal cells, accounting for over 30,000 members [14]; this number is increasing rapidly along with the discovery of different piRNA isoforms [15]. They are known to act specifically in germline cells to preserve the genome integrity from the deleterious effects of transposable elements, through both transcriptional and post-transcriptional repression mechanisms (i.e., suppressing transposon activity, protecting the telomere, and regulating RNA silencing and epigenetic control by establishment of a repressive chromatin state) [16]. piRNAs have more recently been found also in somatic cells [17], where their expression is regulated in a tissue-specific manner, although the impact of this discovery is not yet completely understood [18].

The defective or aberrant expression of sncRNAs has been associated with several human diseases, and the interest in their biological roles has increased concurrently [18,19].

Advances in methods that analyze RNA populations have allowed a rapid quantitative and qualitative characterization of small RNAs at the cellular and/or tissue level. As a result, a huge quantity of "omics" data are currently available, stored in specialized databases and representing a valuable source of information which could potentially provide surprising and useful details on genome architecture and regulatory frameworks [20]. Precious hints indeed derive from the analysis of sncRNA sequences that could reveal the presence of conserved domains plausibly enabling additional RNA–RNA or RNA–DNA interactions, gathering them in the context of specific cellular functions on a genome-wide scale.

In this scenario, the aim of this study was to analyze the complete subset the whole sequences of miRNAs and piRNAs stored in the main databases in order to identify the occurrence of conserved motifs and subsequently predict possible specific interplay with TFs within genome-scale regulatory networks. We propose a direct interaction between the sncRNAs (either miRNAs or piRNAs) with conserved common motifs (transcription factor binding site (TFBS)) on DNA sequences. This original analytical workflow allowed identifying the occurrence of conserved motifs in both miRNAs and piRNAs and categorizing these sncRNAs based on TFBS domains. The data produced by this in silico pipeline point towards the hypothesis that miRNAs and piRNAs share DNA-binding motifs with TFs. We then propose that these sncRNAs may be categorized based on TFBS. Interestingly, the functional annotation of putative target genes allowed evidencing that these binding motifs are specifically enriched in biological networks involved in embryonic development. Indeed, the analysis allowed pointing out that different miRNA and piRNA classes, sorted by TFBS, have differential implication in biological pathways, as they are able to regulate multiple target genes sharing the same conserved TFBS. Our in silico analysis provides original predictive computational data, paving the way to further biological studies that may test novel nuclear roles of miRNAs and piRNAs in gene expression regulation.

2. Materials and Methods

The analysis pipeline was performed using different bioinformatics tools available online in the MEME-suite collection of motif-bases sequence analysis tools (http://meme-suite.org/). The experimental pipeline consisted of three main steps, as schematized in Figure 1.

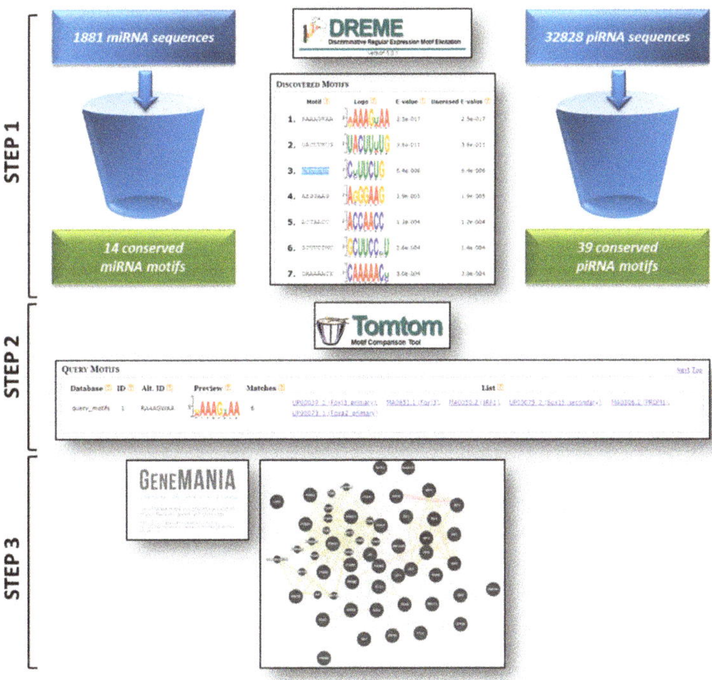

Figure 1. Experimental workflow. A schematic representation of the pipeline integrating the different analytical software is provided. The details of the analytical steps are described in the text (Section 2).

The entire subgroup of *Homo sapiens* mature miRNAs (1881 mature miRNA sequences from http://www.mirbase.org/ftp.shtml; only unique sequences were considered) and piRNAs (32,826 piRNA sequences from http://regulatoryrna.org/database/piRNA/download.html) were analyzed separately.

In the first step, all reliable conserved motifs were categorized by searching in miRNA and piRNA sequences using the Discriminative Regular Expression Motif Elicitation (DREME) tool from MEME-suite (http://meme-suite.org/tools/dreme) [21]. The DREME tool allows finding relatively short motifs (up to eight positions) using sets of sequences (in our case, miRNAs and piRNAs, as reported in * fast sequences) as input. The program does not need a control set since it shuffles the primary set to provide it. Moreover, it exploits Fisher's exact test to determine significance of each motif found in the positive set using a significance threshold. The motifs identified by this approach were stopped when the next motif's E-value threshold exceeded 0.05 (default threshold) [21]. The identified motifs were mapped within the miRNA sequences in order to observe their localization and possible relation with the seed sequences, according to the MicroRNA Target Prediction Database (miRDB; http://www.mirdb.org).

In the second step, all the obtained motifs were used as query for Tomtom, a motif comparison tool within the MEME-suite (http://meme-suite.org/tools/tomtom), which compares the newly identified motifs against a database of known motifs (i.e., JASPAR). JASPAR CORE is a database that contains a curated and nonredundant set of open access data collections of experimentally discovered and proven TF binding sites [22,23]. Tomtom ranked the motifs in the database and produced an alignment for each significant match, searching one or more query motifs against one or more databases of target motifs (and their reverse complements when applicable). The report for each query was a list of target motifs, ranked by *p*-value in the order that the queries appear in the input file. The E-value and the q-value for each match were also reported. The q-value is the minimal false discovery rate at which the observed similarity would be considered significant. Tomtom estimated q-values from all the match *p*-values using the Benjamini and Hochberg method. By default, significance was measured by q-value of the match [22].

For all motif queries, a list of transcription factors that contained the common conserved domain was obtained (all the motifs with corresponding TF lists are reported in Tables 1 and 2).

Table 1. MicroRNA (miRNA) conserved motifs.

Motif Sequence	miRNA-Contained Sequence Motif	miRNA-Contained Complement Sequence Motif	Transcription Factors
RAAAGWAA	100	8	AR, BCL6, ETS2, FOXA1, FOXA3, FOXJ3, FOXK1, FOXM1, FOXP1, FOXQ1, GATA3, IRF1, IRF2, IRF3, IRF7, IRF8, IRF9, LEF1, LHX3, MEF2B, NFATC1, NR2E3, NR4A2, PRDM1, PRDM6, RORG, SMARCA1, SOX2, SOX3, SOX4, SRY, STAT2, ZFP28, ZIM3, ZNF274, ZNF354A, ZNF394, ZNF85
UACUUWUG	59	2	CDX1, FOXA1, FOXA2, FOXA3, FOXC1, FOXM1, LEF1, LHX3, MEF2A, MEF2B, MEF2C, MEF2D, NR2E3, POU5F1, ZNF708
ACCAACC	33	1	ARI5B, FOXI1, GLI3, NFYA, RUNX2, SALL4, Z324A, ZIC1, ZNF449
CYUUCUG	106	33	ATOH1, BCL6, EHF, ERG, ETS1, ETV4, NDF1, NDF2, NR1D1, OLIG2, OSR2, PRGR, SMAD4, STAT2, THA11, ZIC3, ZIM3, ZN143, ZN436, ZN528, ZN547, ZN768, ZNF76, ZSC31
ASGGAAG	60	10	E2F6, ELF1, ELF2, ELK1, ELK4, ETS2, ETV1, FEV, GABPA, IRF3, NFKB1, OLIG2, PAX6, T, ZNF257, ZNF341, ZNF436, ZNF528, ZSCAN31
GCUUCCHU	41	4	ELF3, ERG, ETS2, FEV, FEZF1, FLI1, FOXK1, NHLH1, NFKB1, NFKB2, OLIG2, SMARCA1, SMARCA5, SOX10, ZNF341, ZNF394, ZNF436, ZNF502, ZNF528, ZNF582, ZNF85

Table 1. *Cont.*

Motif Sequence	miRNA-Contained Sequence Motif	miRNA-Contained Complement Sequence Motif	Transcription Factors
CAAAAACY	35	2	HOXA10, NFATC3, RUNX2, ZIM3, ZNF384
UUACBGU	40	4	ELK1, ETS1, ETV4, OVOL1, RFX3, ZIM3
YGGUUUUU	37	3	AIRE, FOXO4, FOXQ1, HXA10, HXA13, MEF2C, NR2E3, RUNX1, RUNX2, RUNX3, TWST1, ZN384
UGUGAY	179	90	ERR1, RXRB, GFI1, GFI1B, MITF, TFEB, JUNB, TFE3, ESR1, RARG, FOXI1, USF1, ZIC3
KAGGUUG	74	21	DUX4, HIC1, ZIM3, ZN136, ZN768
GGKUGGGG	48	9	E2F4, E2F6, EGR1, EGR2, ESR2, GLI3, KLF1, KLF12, KLF15, KLF3, KLF4, KLF5, KLF6, KLF9, MAZ, MXI1, PATZ1, PRDM14, RXRA, SALL4, SP1, SP2, SP3, SP4, SREBF1, SREBF2, TAL1, TBX3, VEZF1, WT1, ZBTB17, ZIC1, ZNF281, ZNF449, ZNF467
GGAMAG	153	78	BCL6, E2F1, E2F4, E2F6, E2F7, ERG, ETS1, ETV4, GATA1, IRF3, MEIS2, MYOD1, MYOG, NFATC1, NFATC2, NFATC3, NFATC4, NFKB1, NFKB2, NR1D1, NR1I3, PBX1, PRDM1, RELB, REST, STAT1, RELA, TFDP1, TGIF1, ZNF257, ZNF274, ZNF335
AUUACUUU	25	1	ALX1, CDX2, DUX4, EVX2, FOXA1, FOXA2, FOXA3, FOXC1, FOXK1, FOXM1, HNF6, IRF1, IRF2, IRF7, IRF8, LHX2, LHX3, NR2E3, PBX2, PRDM1, ZFP28, ZN394, ZNF85

Table 2. PIWI-interacting RNA (piRNA) conserved motifs.

Motif Sequence	piRNA-Contained Sequence Motif	piRNA-Contained Complement Sequence Motif	Transcription Factors
HTTCY	11887	9673	BCL6, E2F1, E2F3, EHF, ELF1, ELF2, ELF3, ELF5, ELK1, ELK4, ERG, ETS1, ETS2, ETV1, ETV2, ETV4, ETV5, FEV, FLI1, GABPA, HSF1, NFATC2, NFATC3, NFKB1, OSR2, PRDM6, SMARCA1, SP4, STAT5A, STAT5B, STAT1, STAT3, STAT4, STAT6, TFDP1, ZFP28, ZNF317, ZNF394, ZNF418, ZNF528, ZNF680
CAYCW	11906	10059	PDX1, CREB1, SNAI1, TBX3, CEBPG, ATF4, SNAI2, REST, ZEB1, NDF1, NDF2
ACTCGYG	345	55	CLOCK
CGTWCCCA	187	9	NFKB1, NFKB2, RFX1, RFX2, RFX3, RBPJ
CACGK	1749	1151	ARNT, ATF3, ATF6A, BHLHE40, BMAL1, CLOCK, EPAS1, HIF1A, MAX, MITF, MTF1, MXI1, MYC, MYCN, TFE3, TFEB, USF1, USF2
GACKCCTC	204	47	BACH2, CRX, FOSB, FOSL1, FOSL2, JUN, JUND, ZNF317
ACCWY	4967	4083	NR2F1, NR2F2, ERR2, ESRRG, GLI3, NR2C1, NR4A1, NR4A2, PPARG, REST, RUNX1, RUNX2, RUNX3, RXRA, RXRG, TBX21, TBX3, ZIC1, ZNF250, ZNF8
AGGTTKGA	162	26	TBX3, PRDM14, ZNF324, ZNF449, TBX21, TEAD1
AAAVTGC	412	215	MAFF, NKX3-1, MAF, IRF1, IRF2, ZNF85, IRF7, PRDM1, BATF3, HNF4A, STAT2, HNF4G, FEZF1, MYB, HIF1A, RORG, ZFP42
GGTTCCGA	56	2	NR2C2
GCAGAYAC	143	43	CLOCK, SMAD2, ZNF708, MAFB, FOXM1, MXI1, ZFP42, ZNF547

Table 2. *Cont.*

Motif Sequence	piRNA-Contained Sequence Motif	piRNA-Contained Complement Sequence Motif	Transcription Factors
AGCTSCTG	250	111	BHLHA15, MYOG, TCF3, MYOD1, MYF6, HTF4, ITF2, ATOH1, ZBTB18, ASCL1, OSR2, PTF1A, NFE2L2, NHLH1, MAF, RFX5, ZNF563, CEBPG, ATF4, LYL1, ZIC3, NEUROD2, MAFG, OLIG2, TFAP4
CACTTAGS	116	30	NKX3-1, ISL1, NKX3-2, DLX3, FOXA2, PRRX2, BACH2, NFE2, NOBOX, BACH1, MYC, FOXA1, NFE2L2, MYCN, ARNT
HATCCTA	290	147	ZNF586, CRX, ZNF324, ZNF274
AGGYAG	1145	843	ZNF335, ZNF490, MYB, RFX5, ZNF770, ETS1, ERG, FLI1, SMAD3, ZNF257, ZNF549, PTF1A, ZNF563, ETV2
CCAAAK	911	649	HNF4A, HNF4G, FOXA1, ISL1
ATGAACTC	70	13	NR1I3, NR1I2, NR2C1, VDR, ATF2, ZNF18, RARA, ZNF549, RXRB, NR6A1, ZNF354A, NR1H4
CKGCTAAA	67	12	T, ZNF322
BCATTTC	217	105	BCL11A, EHF, ELF1, ELF2, ELF3, ELF5, ELK1, ELK4, ERG, ETS1, ETV1, ETV2, ETV4, ETV5, FEZF1, GABPA, MAFF, NFKB1, POU2F1, POU2F2, POU3F1, POU3F2, REL, SPI1, SPIB, STAT5B, STAT1, STAT2, STAT3, RELA, THRA, ZNF354A, ZFP42, ZFP82, ZNF140, ZNF528
TAAGGGTA	48	5	ZNF264, NKX3-2, CRX, ZNF667, DUX4
GTACGWCA	45	4	EPAS1, ARNT, CREB1, ATF1, RORG, HIF1A, CREM, BHE40, ATF3
CGRTGCCC	44	4	HIC1, NFIA, CUX1
GAACGGGY	41	3	HIF1A, ZBT14, ZNF18
CVTGGA	1343	1057	ZNF436, TEAD4, BCL6, RARG, REST, TEAD1, STAT1, STAT5A, STAT3, ZBTB6, STAT4, SNAI1, POU3F2, TP63, ESRRG
GTAGCTAS	55	9	RFX5, BHLHA15
ATCGCTGA	47	6	NFE2L2, MAFK, MAFG, OZF, ZFP42, ZNF335, ZNF528
ABGTTTA	124	50	FOXC1, FOXJ3, FOXA3, FOXK1, FOXO4, FOXQ1, FOXA1, FOXA2, MEF2D, HNF1B, FOXJ2, FOXP2, HNF1A, FOXO3, SRY, MEF2C, MEF2A, FOXP1, MEF2B, FOXO1, FOXM1
CCAMTAAC	60	14	CREM, ETS1, FOXI1, HOXA13, HOXB13, HOXB4, ISL1, RUNX2, VDR
KGGCTTA	219	120	ZNF528, ZNF41, OTX2, ZNF449, ZNF214
GTGTYTA	202	108	CLOCK, FOXA1, FOXA2, FOXJ2, FOXJ3, FOXK1, FOXO3, FOXO4, FOXP1, FOXP2, FOXQ1, MEF2B, NKX3-1, PRDM14
ATTGCACG	23	//	CEBPA, CEBPD, CEBPB, ATF4, CEBPG
CCTARAG	186	99	BCL6, ZNF436, STAT5A
AKGAGGAC	108	46	ZNF816, ZNF586, THRB, ELF5, ZKSCAN1, CRX, ATF2, ZNF263, ETV5, ZIC3, SPI1, RXRB
AAGGSCAC	80	29	ZNF667, ZNF264, NR4A2, ERR1, NR4A1, STF1, RXRG, NR5A2, TFAP2C, ERR2, ZNF214, ESRRG, HNF4G, SOX9, TFAP2A, NR1H3, RXRB, HNF4A
RATGGAA	87	34	ZNF502, SMARCA5, ZFP82, ZNF582, ZNF394, POU2F1, NFATC1, ZNF264, ZNF354A, FOXK1, ZFP28, HOXA1, BATF3, POU3F1, ISL1, ZNF8, ZNF85, TWIST1, STAT2, NFKB1, IRF3, SOX2, NFATC3, ATF4, NFATC2, PRDM6, CEBPG, ETS2
ACTGWTCG	34	5	ZNF691, CUX1
GAACWCA	338	226	NR3C1, VDR, RXRB, PGR, SOX4, ZKSCAN1, FOXP1
CCGTAGCY	29	3	MYOG, RFX1, RFX2, MYCN
ACCDACTG	104	45	ZKSCAN1, MYB, NEUROD1, SNAI1, KLF8, NEUROD2, ATOH1, TFAP4, HTF4, BHLHA15, NR1D1

Finally, in the third step, all lists of putative miRNAs and piRNAs that contained motif-related TFs were individually loaded in GeneMANIA (https://genemania.org/). This is a flexible, user-friendly web interface for generating hypotheses about function, through analyzing gene lists and prioritizing them based on literature-proved biological functions. GeneMANIA allowed clustering functionally related TFs, using available genomic and proteomic data (protein–protein, protein–DNA interactions, signaling pathways, protein domains, and phenotypic screening profiles) introducing weights that indicate the predictive value of each selected dataset for the query [24].

The lists of annotated functions were then used for the biological interpretation to achieve a functional hypothesis. In particular, using a Benjamini–Hochberg false discovery ratio (FDR) multiple testing correction (also known as the 'q-value') associated with each 'function', we selected for the analysis only GO terms with $FDR \leq 0.05$, in order to reduce redundancies.

3. Results

3.1. Conserved Motif Annotation in miRNAs and piRNAs

The DREAM tool (step 1 of our workflow, see Figure 1) allowed identifying conserved motifs in both miRNAs and piRNAs. These motifs were predicted to bind specific DNA sequences on the basis of common domains with TFs, suggesting a putative involvement of sncRNAs in nuclear gene regulation.

In particular, we identified conserved motifs in 66.7% of mature miRNAs and in about 94% of the piRNAs analyzed. Particularly, 14 and 39 conserved motifs were identified in miRNAs and in piRNAs, respectively (Tables 1 and 2). The analysis was performed considering complementary sequences as well, in order to identify motifs able to bind on both directions and on either DNA strand.

When mapping the precise location of the motifs within the miRNA sequences, we found that these do not necessarily correspond to the seed sequence positions. In some cases, the sequence motif is localized at the 5' end of the tested miRNA, where it only partially overlaps with the seed sequence. In all other cases, the motifs were mapped in variable regions of the mature miRNA sequences.

Of the 14 conserved motifs identified in miRNAs, four motifs (RAAAGWAA, CYUUCUG, UGUGAY, and GGAMAG) were present in more than 100 miRNAs when also considering complementary sequences. The least represented motif is AUUACUUU, which can be recognized in as few as 26 miRNAs (entire range: from 26 to 269 miRNAs; see Table 1).

The number of piRNAs containing one or more of the 39 putative DNA-binding motif ranges from 23 for the ATTGCACG motif to 21,965 for the CAYCW domain (Table 2). It is noteworthy that, in some piRNA sequences, more than one conserved motif has been identified, hence the extent of DNA-binding partners and corresponding target genes would be even greater.

The Tomtom tool (step 2 of our workflow, see Figure 1) allowed the annotation of the target TFs containing the DNA-binding motifs found in the tested miRNAs (Table 1) and piRNAs (Table 2). Our data showed that while the number of putative transcription factors sharing either of the conserved DNA-binding motifs identified in miRNAs is relevant (Table 1), some of the conserved motifs (namely ACTCGYG and GGTTCCGA) identified in piRNAs could be associated to just one gene (Table 2).

3.2. Computational Pathway Analysis of TFs

Through the GeneMANIA tool (step 3 of our workflow, see Figure 1) we were able to predict the functional implication of the putative TFs sharing the motifs identified in miRNAs and piRNAs (listed in Tables 1 and 2). This allowed speculating on their functional implication and networking in biological processes. The entire lists of TF-related 'functions' computed through the GeneMANIA software are provided in Supplemental File S1 (miRNA and piRNA motif-associated functions). In order to rationalize the wide range of GO terms associated as 'functions' to each TF list, we performed a further additional clustering of similar functional categories by grouping biologically related functions shared by TF lists related to both miRNA and piRNA motifs. Overall, we categorized 644 functions with corresponding GO terms into 21 functional clusters, listed in the Supplemental File S1. Specifically,

12 of 14 miRNA motifs were linked by GeneMANIA to annotated functions, and only seven of these domains could be positively associated within 16 functional clusters, as schematized in Figure 2 (see also Supplemental File S1). Also, 35 of 39 piRNA motifs were associated with biological functions; of these, 29 domains could be arranged in 21 clusters, with five new clusters in addition to those identified for miRNA domains (namely 'Apoptosis and cellular response to stress stimuli', 'Endocrine signaling', 'Metabolic processes', 'Regulation of circadian rhythm', and 'RNA-mediated gene silencing') (Figure 3, Supplemental File S1).

As expected, nonspecific "DNA-binding" activities, inherently associated to all TFs, accounted for 88 out of the 644 (14%) annotated functions (grouped in the 'DNA interaction/gene expression regulation' functional cluster, Supplemental File S1). In particular, these were annotated in all but one (namely GCUUCCHU) of the miRNA motifs and in all but four of the piRNA motifs (namely HATCCTA, TAAGGGTA, GAACGGGY, KGGCTTAA). Furthermore, 118 out of the 644 (18%) GO annotations found for most miRNA and piRNA motifs represented general biological processes and pleiotropic signal transduction pathways that cannot be clearly categorized into univocal biological functions (grouped in the 'General cellular processes/pleiotropic signal transduction pathways' functional cluster, Supplemental File S1). Due to the nonspecificity of these annotations, the above specified clusters were not considered further in the biological interpretation.

Interestingly, after excluding nonspecific functions, as many as 293 functions out of the 644 (45%) GO annotations were associated with specific developmental stages, from early embryo formation to detailed organogenetic paths (Supplemental File S1). We have further described these functional categories and the related sncRNA motifs, in the attempt to provide an experimental hypothesis to be tested in wet lab functional studies (Figures 2 and 3).

In particular, the analysis of the motifs contained in miRNAs (Supplemental File S1) allowed hypothesizing a possible implication of selected motifs in the regulation of specific developmental stages, according to a specific timeframe (Figure 2). In particular, for example, the miRNA sequence motifs RAAAGWAA, UACUUWUG, YGGUUUUU, and AUUACUUU apparently represent putative DNA-binding domains shared not only by TFs involved in the regulation of early developmental stages and morphogenetic events ('Embryonic development, early stages up to gastrulation' functional cluster), but also with those involved later events, such as neural, urogenital and skeletal system development or miscellaneous developmental processes (Figure 2). On the contrary, the late stages of embryonic development ('Embryonic development, late stages, tissue patterning') seemed to be specifically regulated by the RAAAGWAA motif (Figure 2). Hemopoiesis and endocrine and respiratory system development functions also appeared enriched among the GO terms for the RAAAGWAA motif, while UACUUWUG, ACCAACC, and AUUACUUU domains were found to be involved in the regulation of digestive system development and RAAAGWAA, UACUUWUG, and YGGUUUUU motifs were related to cardiovascular system and muscle development functions (Figure 2). In addition, our in silico analysis showed that RAAAGWAA, GCUUCCHU, GGAMAG, and AUUACUUU binding motifs could affect the expression of genes associated with immune system development and function (Figure 2).

The functional cluster 'Stem cell homeostasis and differentiation' (Supplemental File S1) was not considered further in the biological interpretation of miRNA motifs, since this function cannot be associated with a specific timeframe but rather is part of all stages of prenatal and postnatal tissue development and homeostasis.

Moreover, our analytical workflow also provided evidence of a spatial and temporal distribution for TFs sharing DNA-binding motifs with piRNAs, revealing an intricate network of multiple connections with redundant functions (Figure 3). As already mentioned, the computational analysis of piRNA motifs displayed the enrichment of additional functional clusters, beyond that associated with embryo development, and the regulation of these functions could be thus specifically connected with this class of sncRNAs. Moreover, the gene regulation connected to specific biological processes seemed to be very specific for some types of sequence domains compared with those identified in miRNAs. In further detail, each functional cluster displayed in Figure 3 seemed to be regulated by a characteristic set of DNA-binding domains: if on the one hand it was possible to identify functional clusters regulated by over six motifs (namely 'Apoptosis and cellular response to stress stimuli', 'Immune system development and function', and 'Metabolic processes' functional clusters), other groups appeared to be peculiarly associated with only a single motif (AAAVTGC with hemopoiesis, ABGTTTA with respiratory system development, GCAGAYAC with RNA-mediated gene silencing) (Figure 3). Interestingly, the implication of this piRNA motif in functions related to RNA-interfering mechanisms could directly support the interplay with the annotated TFs and sncRNAs in gene regulation. In addition, for example, some domains were found to be involved in the regulation of circadian rhythm (ACTCGYG, CACGK, ACCWY, ATGAACTC, and GTACGWCA), while others were involved in the regulation of in stem cell homeostasis and differentiation (GCAGAYAC, CCAAAK, CKGCTAAA, and ABGTTTA) (Figure 3).

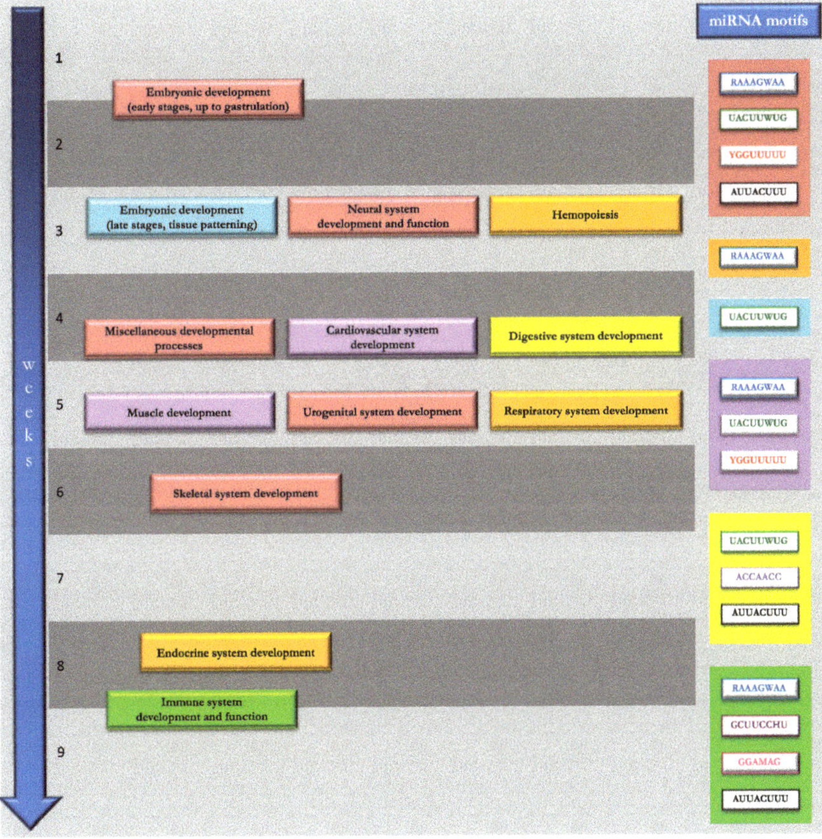

Figure 2. Schematic representation of spatial and temporal distribution of putative DNA-binding motifs identified in microRNAs (miRNAs). The functional clusters identified with the computational pathway analysis of transcription factors (TFs) sharing the motifs identified in miRNAs are depicted as colored boxes distributed along the timeline (arrow on the left side) of human embryo development. The colored boxes (right side; same color ID scale) group the motifs identified in miRNAs for which developmental functions were enriched, according to GeneMANIA functional interpretation (see text for details).

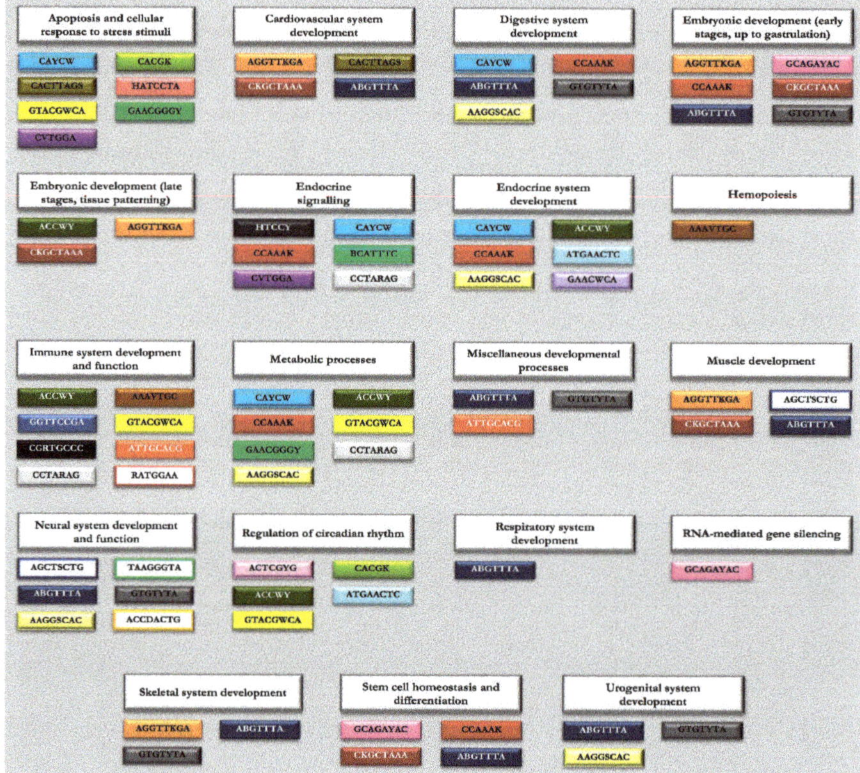

Figure 3. Schematic representation of the enriched functions for PIWI-interacting RNA (piRNA) motifs. The scheme provides an overview of the motifs associated with each functional cluster derived from the computational pathway analysis (based on GeneMANIA tool) of transcription factors (TFs) sharing the same domains with piRNAs (see text for details). A specific colored box is assigned to each piRNA motif.

4. Discussion and Conclusions

The complete understanding of the complex miRNA-mediated regulatory network in cells and organisms has not been achieved yet. The selection of molecular interacting targets by miRNA has been long considered to be primarily dictated by sequences at their 5′ end (nucleotides 2 to 7, known as "seed" sequence). Nonetheless, distinct studies have suggested that miRNAs contain additional sequence elements that control their posttranscriptional behavior, including their subcellular localization.

Indeed, it is currently confirmed that mature miRNAs reside in the nucleus, where they participate at several levels of gene expression regulation [9–11]. Different pieces of evidence pointed out that RNA-induced silencing complex (RISC) protein complexes also exist in the nucleus, where they actively contribute to the nuclear import of miRNAs (Figure 4) [25–27].

The expression enrichment of different miRNA sets in the nucleus seems to vary based on cell type, function, and activity status, or in response to environmental stimuli [12,28]. Numerous efforts have been devoted to the identification of sequence regions within miRNAs able to affect and direct their nuclear import. Hwang and co-workers reported that a hexanucleotide element (AGUGUU) at the 3′ end may affect the subcellular localization of mature miRNAs. This sequence motif apparently acts as a nuclear localization signal enabling the nuclear import of mature miRNA from the cytoplasm [29]. Interestingly this motif resembles the reverse sequence of the UGUGAY motif identified in this study.

Also, another study found that two additional sequence motifs were found in nuclear-localized miRNAs expressed by endothelial and muscle cells upon hypoxic conditions [28]. Moreover, another study showed that most of the nucleus-enriched miRNAs share a common sequence motif with homology to the consensus MYC-associated zinc finger protein (MAZ) transcription factor binding element [30].

The noncanonical nuclear role of miRNAs in the regulation of gene expression at the transcriptional level is yet to be fully clarified. Alternative mechanisms have been proposed to date that are not necessarily mutually exclusive but rather suggest that miRNAs may intervene at several levels in the gene expression regulatory network occurring in the nucleus. Most of the described mechanisms members of the nuclear subfamily of argonaute (Ago) proteins, key components of RISC complexes, as key mediators. Such nuclear miRISC complexes may bind long noncoding RNAs (lncRNAs) by sequence complementarity and modulate their function [11,12]. The lncRNA class of ncRNA includes epigenetic mediators acting in the nucleus (including 'promoter-associated' and 'enhancer-associated' RNAs and 'gene body-associated' RNAs). These are in turn able to influence chromatin organization, acting as structural scaffolds of nuclear domains, and to mediate transcriptional/cotranscriptional regulation [31]. Other experimental studies suggested miRNAs to be involved with the ribogenesis process occurring in the nucleolus, while others described their participation in the regulation of alternative splicing (see [11] for a review).

Converging evidence also showed that miRNAs may modify (either activate or suppress) gene transcription by interacting with chromatin, besides acting at the post-transcriptional level in the cytoplasm [9,32]. Mature miRNAs in the nucleus may indeed directly bind double-stranded DNA within specific target sequences [12,33]. Specifically, it has been reported that miRNAs can form triple-helical structures with specific regions of DNA through either Hoogsteen or reverse Hoogsteen pairings [33]. Nonetheless, the likelihood of their effective occurrence of such pairing modalities in physiological conditions is still widely debated [34,35]. It is instead more likely that miRNAs regulate gene transcription by binding to promoter sequences in an Ago-dependent manner, as demonstrated in a number of studies [32,36–38]. Ago proteins are known to act in the nucleus, despite their structure not including a known DNA-binding domain; therefore, their interactions with chromatin and chromosomes might be mediated by miRNAs. In particular, the nuclear Ago1 protein directly interacts with RNA polymerase II (RNAPII) and is preferentially enriched in promoters of transcriptionally active genes [39]. The Ago1–RNAPII interaction decreases if miRNAs are depleted, hence suggesting that Ago1–chromosomal interaction is mediated by miRNAs [39].

Alternatively, miRNAs may recognize complementary sequences on nascent RNAs, in a cotranscriptional mechanism, forming double-stranded RNAs that determine the recruitment of protein complexes able to modify chromatin accessibility and thus RNAs levels [11,12].

Finally, a third model has been also proposed, according to which miRNA–Ago complexes directly target one of the DNA strands when the target promoter region is in an open configuration during the transcription initiation process [12].

Although numerous studies have indicated that the seed sequence can also mediate the recognition of miRNAs' nuclear targets, a recent study supports a model in which a miRNA can form a hybrid with promoter region to modulate transcription through its nonseed region [40]. In this scenario, the interplay between miRNAs and TFs has been emerging as a key mechanism within the complex network of transcriptional regulatory networks occurring in the nucleus. Such interactions are believed to rely on the presence of conserved regulatory motifs and are needed to finely tune developmental programs in multicellular organisms [41]. Hence, with the aim of making our results as extensive as possible, we decided to include in our in silico analysis all the sequence motifs independently by the overlapping with the seed sequence. Since the mechanisms of miRNA-mediated gene regulation have not been completely clarified and new models continue to be identified in different experimental conditions, our approach aims to avoid losing some important data.

A similar interplay with TFs has not been explored in piRNAs to date, even though increasing roles for this class of sncRNAs, further preserving genome integrity in germline cells, have been recently recognized. The mechanisms at the base of piRNA biogenesis and function have become increasingly clear (Figure 5), and growing evidence suggests that specific piRNA expression patterns can be recognized in pathological conditions, including cancers [42,43].

Understanding the crosstalk between sncRNAs and TFs at a cotranscriptional level could provide new clues towards the involvement of miRNAs and piRNAs in the control of specific events of gene expression regulation in the nucleus during development in humans.

The computational workflow exploited in this study allowed posing an experimental hypothesis according to which conserved and recurrent sequence motifs found in miRNAs and piRNAs, complementary to transcription factor binding sites (TFBS), could influence the bond and activity of TFs on the same target genes. Our results enabled categorizing different classes of motifs, associated to TFs that have known biological roles, hence predicting the possible biological consequence of the putative miRNA nuclear localization and function.

The biological interpretation of the enriched functional terms in TFs indeed allowed categorizing miRNA motifs according to their involvement in key steps of the human developmental path, suggesting that different miRNA profiles exist in different developmental stages and vary in their nuclear expression across different tissue types. The direct competition/collaboration with TFs that our data suggest might provide a finer regulatory control and could explain the prompt canalization of genetic programs to maintain and stabilize the phenotypic reproducibility of embryogenesis. This type of interaction could increase the speed and efficiency of response of embryonic cells exposed to continuous differentiation stimuli. Indeed, the established mechanisms of miRNA-mediated expression regulation, based on their biding on the 3′-UTR of target mRNAs in the cytoplasm, inevitably occurs at the post-transcriptional level, while the 5′ end DNA-binding event proposed in this model occurs during or right before transcription [10–12]. This can therefore cooperate with several mechanisms to accomplish the finely tuned regulatory network especially needed during early developmental stages.

The results obtained in the in silico analysis of piRNA motifs yielded original data providing a model for additional roles for piRNAs in somatic cells, to be further explored in the wet lab. Indeed, piRNA motifs were also found to be associated to TFs with reproducible functions exerted in somatic tissues, including tissue-specific metabolic pathways. It has been shown that the PIWI–piRNA complex binds its genomic target in euchromatin through a nascent transcript and, in heterochromatin, predominantly through a direct piRNA–DNA interaction [44]. Although the information on piRNA roles in somatic tissues is still limited, these data may contribute to postulating new functional roles, regulatory functions, and towards their translation into the identification of new markers of biological processes and/or diseases.

These data may also suggest a feasible way to categorize functional piRNA subclasses distributed in different tissues, on the basis of the presence of conserved motifs, that could reflect their roles in shared regulatory networks and/or developmental timeframes. Each of these families could intervene in crucial parts of the epigenetic control, maintaining genomic integrity, repressing the mobilization of transposable elements, and regulating the expression of downstream target genes via transcriptional or post-transcriptional mechanisms as already reported by two independent research teams in studies of model organisms [45,46]. These groups independently observed that parental responses to the environment are passed to offspring by small RNAs, suggesting that even environment-related behavioral traits can be passed down through generations by transgenerational epigenetic inheritance (TEI), even though the underlying mechanisms are unclear [45,46].

Our observations, though still preliminary, could propose novel testing hypotheses to be investigated in a biological system, towards the clarification of novel aspects of sncRNA-based epigenetic regulation of cellular functions at the organism level.

Further in vitro analyses will be necessary to support at the functional level the evidence derived from this in silico approach. In particular, in-depth in vitro studies at the genome-wide level will be needed to delve into the subcellular location of each class of motif-grouped sncRNAs and to clarify their involvement in the predicted biological pathways.

The extended knowledge of these novel sncRNA mechanisms of action has a deep translational relevance, considering their extensive application in "theranostics": a differential expression analysis of these sncRNA sequence motifs could enable identifying tissue- or organ-specific biomarkers of pathway function/dysfunction. On the other hand, targeting RNA metabolism is being exploited as a strategy to recover RNA alterations in a variety of diseases, paving the way to RNA-based therapeutic strategies [19].

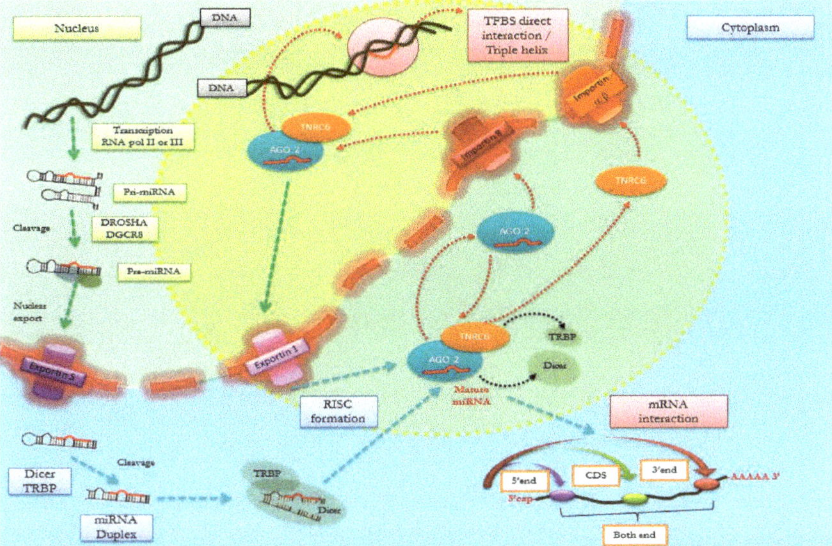

Figure 4. Several steps of microRNA (miRNA) biogenesis and nucleus–cytoplasm transport. Mature miRNAs derive from longer double-stranded primary transcripts (pri-miRNA), which are recognized and processed in the nucleus by the Drosha protein/DGCR8 microprocessor complex subunit (DGCR8) complex into shorter precursors folded in a hairpin loop structure (pre-miRNA). Pre-miRNAs are then exported to the cytoplasm (through Exportin 5) where they are first cleaved by Dicer and later processed by RNA-induced silencing complex (RISC) to form the mature miRNAs. Transactivation response element RNA-binding protein (TRBP) intervenes in the stabilization of Dicer. RISC, which includes Protein argonaute-2 (Ago2), also participates in the identification of miRNAs' targets. The integrative miRNA network highlights, in the yellow circle, the import of Ago2 with mature miRNA into the nucleus via Importin 8 and trinucleotide repeat-containing gene 6A protein (TNRC6), another component of RISC complex, via Importin β. Nuclear RISC is again assembled to elicit pleiotropic effects by regulating multiple pathways with a direct interaction on DNA transcription factor binding sites (TFBSs) and possible formation of triple-helix structures.

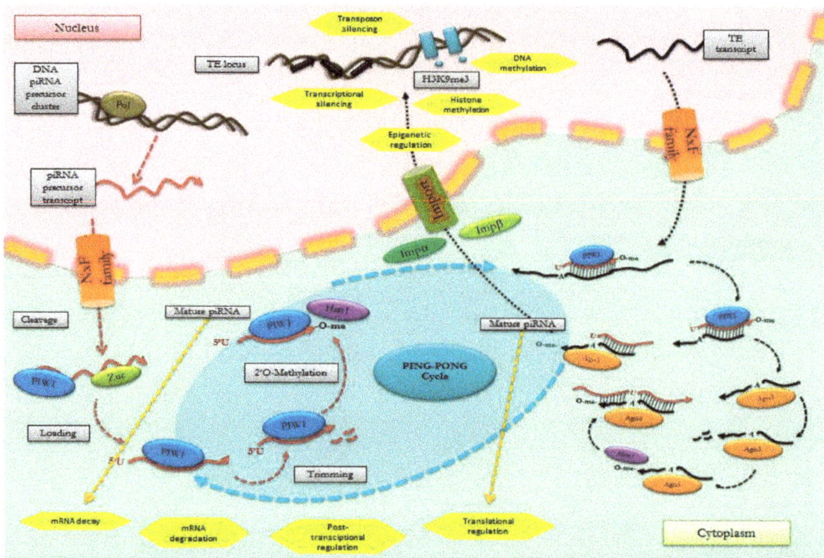

Figure 5. Different stages of PIWI-interacting RNA (piRNA) biogenesis and function. Mature piRNAs are derived from precursor RNAs following a post-transcriptional processing through two alternative mechanisms. The primary maturation pathway involves cleavage of long, single-stranded piRNA clusters and the binding with PIWI proteins in the cytoplasm. The second mechanism is a self-amplifying loop (termed "ping-pong" cycle), in which an antisense piRNA binds PIWI proteins and triggers production of a sense piRNA that binds to Protein argonaute-3 (Ago 3). Nuclear PIWI/piRNA complexes regulate gene and transposon expression by epigenetic modifications. Once the piRNAs are loaded onto PIWI, the activity and/or expression of DNA methyltransferases (Hen 1) is increased, promoting methylation of promoter regions, preventing transcription factor binding, and interacting with histone methyltransferase. Cytoplasmic mature piRNA promotes mRNA decay by interacting with deadenylation complex, inhibits translation by directly binding with translation factors, and modulates cellular signaling by directly regulating the post-translational modifications.

Supplementary Materials: The following are available online at http://www.mdpi.com/2073-4425/11/5/482/s1, Supplemental File S1, Functional Clusters of microRNA (miRNA) and PIWI-interacting RNA (piRNA) domains.

Author Contributions: Conceptualization, M.C. and L.D.P.; Methodology, M.C. and L.D.P.; Software, M.C.; validation, L.D.P., W.L.; Formal analysis, M.C. and L.D.P.; Investigation, M.C. and L.D.P.; Resources, W.L. and N.B.; Data curation, M.C., L.D.P., N.B. and W.L.; Writing—original draft preparation, M.C. and N.B.; Writing—review and editing, N.B. and W.L.; Visualization, M.C. and L.D.P.; Supervision, N.B. and W.L.; Project administration, W.L.; Funding acquisition, W.L. All authors have read and agreed to the published version of the manuscript.

Funding: This research was supported by the Università Cattolica del Sacro Cuore [Linea D.1–2018 to W.L.].

Acknowledgments: The authors are grateful to Carol Basilio for language editing and proofreading.

Conflicts of Interest: The authors declare that they have no competing financial and/or non-financial interests, or other interests that might be perceived to influence the results and/or discussion reported in this paper.

References

1. Richard Boland, C. Non-coding RNA: It's Not Junk. *Dig. Dis. Sci.* **2017**, *62*, 1107–1109. [CrossRef] [PubMed]
2. Diamantopoulos, M.A.; Tsiakanikas, P.; Scorilas, A. Non-coding RNAs: The riddle of the transcriptome and their perspectives in cancer. *Ann. Transl. Med.* **2018**, *6*, 241. [CrossRef] [PubMed]
3. Holley, R.W.; Apgar, J.; Everett, G.A.; Madison, J.T.; Marquisee, M.; Merrill, S.H.; Penswick, J.R.; Zamir, A. Structure of a ribonucleic acid. *Science* **1965**, *147*, 1462–1465. [CrossRef] [PubMed]

4. Gilbert, W. The RNA world Superlattices point ahead. *Nature* **1986**, *319*, 618. [CrossRef]

5. Martinez, N.J.; Walhout, A.J.M. The interplay between transcription factors and microRNAs in genome-scale regulatory networks. *BioEssays* **2009**, *31*, 435–445. [CrossRef]

6. Kim, V.N.; Nam, J.W. Genomics of microRNA. *Trends Genet.* **2006**, *22*, 165–173. [CrossRef]

7. Liu, B.; Li, J.; Cairns, M.J. Identifying miRNAs, targets and functions. *Brief. Bioinform.* **2014**, *15*, 1–19. [CrossRef]

8. Kozomara, A.; Griffiths-Jones, S. MiRBase: Integrating microRNA annotation and deep-sequencing data. *Nucleic Acids Res.* **2011**, *39*, 152–157. [CrossRef]

9. Place, R.F.; Li, L.C.; Pookot, D.; Noonan, E.J.; Dahiya, R. MicroRNA-373 induces expression of genes with complementary promoter sequences. *Proc. Natl. Acad. Sci. USA* **2008**, *105*, 1608–1613. [CrossRef]

10. Ni, W.J.; Leng, X.M. Dynamic miRNA-mRNA paradigms: New faces of miRNAs. *Biochem. Biophys. Rep.* **2015**, *4*, 337–341. [CrossRef]

11. Catalanotto, C.; Cogoni, C.; Zardo, G. MicroRNA in control of gene expression: An overview of nuclear functions. *Int. J. Mol. Sci.* **2016**, *17*, 1712. [CrossRef] [PubMed]

12. Liu, H.; Lei, C.; He, Q.; Pan, Z.; Xiao, D.; Tao, Y. Nuclear functions of mammalian MicroRNAs in gene regulation, immunity and cancer. *Mol. Cancer* **2018**, *17*, 64. [CrossRef] [PubMed]

13. Aravin, A.A.; Naumova, N.M.; Tulin, A.V.; Vagin, V.V.; Rozovsky, Y.M.; Gvozdev, V.A. Double-stranded RNA-mediated silencing of genomic tandem repeats and transposable elements in the D. melanogaster germline. *Curr. Biol.* **2001**, *11*, 1017–1027. [CrossRef]

14. Peng, J.C.; Lin, H. Beyond transposons: The epigenetic and somatic functions of the Piwi-piRNA mechanism. *Curr. Opin. Cell Biol.* **2013**, *25*, 190–194. [CrossRef] [PubMed]

15. Zhang, H.; Ali, A.; Gao, J.; Ban, R.; Jiang, X.; Zhang, Y.; Shi, Q. IsopiRBank: A research resource for tracking piRNA isoforms. *Database* **2018**, *2018*. [CrossRef]

16. Samji, T. PIWI, piRNAs, and germline stem cells: What's the link? *Yale J. Biol. Med.* **2009**, *82*, 121–124.

17. Zhang, X.; He, X.; Liu, C.; Liu, J.; Hu, Q.; Pan, T.; Duan, X.; Liu, B.; Zhang, Y.; Chen, J.; et al. IL-4 Inhibits the Biogenesis of an Epigenetically Suppressive PIWI-Interacting RNA To Upregulate CD1a Molecules on Monocytes/Dendritic Cells. *J. Immunol.* **2016**, *196*, 1591–1603. [CrossRef]

18. Yu, Y.; Xiao, J.; Hann, S.S. The emerging roles of PIWI-interacting RNA in human cancers. *Cancer Manag. Res.* **2019**, *11*, 5895–5909. [CrossRef]

19. Buratti, E.; Bhardwaj, A. Editorial: Role of RNA Modification in Disease. *Front. Genet.* **2019**, *10*, 920. [CrossRef]

20. Wang, Z.; Gerstein, M.; Snyder, M. RNA-Seq: A revolutionary tool for transcriptomics. *Nat. Rev. Genet.* **2009**, *10*, 57–63. [CrossRef]

21. Bailey, T.L. DREME: Motif discovery in transcription factor ChIP-seq data. *Bioinformatics* **2011**, *27*, 1653–1659. [CrossRef] [PubMed]

22. Gupta, S.; Stamatoyannopoulos, J.A.; Bailey, T.L.; Noble, W.S. Quantifying similarity between motifs. *Genome Biol.* **2007**, *8*, R24. [CrossRef] [PubMed]

23. Khan, A.; Fornes, O.; Stigliani, A.; Gheorghe, M.; Castro-Mondragon, J.A.; Van Der Lee, R.; Bessy, A.; Chèneby, J.; Kulkarni, S.R.; Tan, G.; et al. JASPAR 2018: Update of the open-access database of transcription factor binding profiles and its web framework. *Nucleic Acids Res.* **2018**, *46*, D260–D266. [CrossRef] [PubMed]

24. Warde-Farley, D.; Donaldson, S.L.; Comes, O.; Zuberi, K.; Badrawi, R.; Chao, P.; Franz, M.; Grouios, C.; Kazi, F.; Lopes, C.T.; et al. The GeneMANIA prediction server: Biological network integration for gene prioritization and predicting gene function. *Nucleic Acids Res.* **2010**, *38*, 214–220. [CrossRef] [PubMed]

25. Rüdel, S.; Flatley, A.; Weinmann, L.; Kremmer, E.; Meister, G. A multifunctional human Argonaute2-specific monoclonal antibody. *RNA* **2008**, *14*, 1244–1253. [CrossRef] [PubMed]

26. Janowski, B.A.; Huffman, K.E.; Schwartz, J.C.; Ram, R.; Nordsell, R.; Shames, D.S.; Minna, J.D.; Corey, D.R. Involvement of AGO1 and AGO2 in mammalian transcriptional silencing. *Nat. Struct. Mol. Biol.* **2006**, *13*, 787–792. [CrossRef]

27. Hicks, J.A.; Li, L.; Matsui, M.; Chu, Y.; Volkov, O.; Johnson, K.C.; Corey, D.R. Human GW182 Paralogs Are the Central Organizers for RNA-Mediated Control of Transcription. *Cell Rep.* **2017**, *20*, 1543–1552. [CrossRef]

28. Turunen, T.A.; Roberts, T.C.; Laitinen, P.; Väänänen, M.A.; Korhonen, P.; Malm, T.; Ylä-Herttuala, S.; Turunen, M.P. Changes in nuclear and cytoplasmic microRNA distribution in response to hypoxic stress. *Sci. Rep.* **2019**, *9*, 10332. [CrossRef]

29. Hwang, H.W.; Wentzel, E.A.; Mendell, J.T. A hexanucleotide element directs microRNA nuclear import. *Science* **2007**, *315*, 97–100. [CrossRef]

30. Goldie, B.J.; Fitzsimmons, C.; Weidenhofer, J.; Atkins, J.R.; Wang, D.O.; Cairns, M.J. MiRNA enriched in human neuroblast nuclei bind the MAZ transcription factor and their precursors contain the MAZ consensus motif. *Front. Mol. Neurosci.* **2017**, *10*, 259. [CrossRef]

31. Sun, Q.; Hao, Q.; Prasanth, K.V. Nuclear Long Noncoding RNAs: Key Regulators of Gene Expression. *Trends Genet.* **2018**, *34*, 142–157. [CrossRef] [PubMed]

32. Huang, V.; Place, R.F.; Portnoy, V.; Wang, J.; Qi, Z.; Jia, Z.; Yu, A.; Shuman, M.; Yu, J.; Li, L.C. Upregulation of Cyclin B1 by miRNA and its implications in cancer. *Nucleic Acids Res.* **2012**, *40*, 1695–1707. [CrossRef] [PubMed]

33. Paugh, S.W.; Coss, D.R.; Bao, J.; Laudermilk, L.T.; Grace, C.R.; Ferreira, A.M.; Waddell, M.B.; Ridout, G.; Naeve, D.; Leuze, M.; et al. MicroRNAs Form Triplexes with Double Stranded DNA at Sequence-Specific Binding Sites; a Eukaryotic Mechanism via which microRNAs Could Directly Alter Gene Expression. *PLoS Comput. Biol.* **2016**, *12*, e1004744. [CrossRef] [PubMed]

34. Kuo, C.C.; Hänzelmann, S.; Sentürk Cetin, N.; Frank, S.; Zajzon, B.; Derks, J.P.; Akhade, V.S.; Ahuja, G.; Kanduri, C.; Grummt, I.; et al. Detection of RNA-DNA binding sites in long noncoding RNAs. *Nucleic Acids Res.* **2019**, *47*, e32. [CrossRef] [PubMed]

35. Zhou, Z.; Giles, K.E.; Felsenfeld, G. DNA-RNA triple helix formation can function as a cis-acting regulatory mechanism at the human β-globin locus. *Proc. Natl. Acad. Sci. USA* **2019**, *116*, 6130–6139. [CrossRef] [PubMed]

36. Kim, D.H.; Villeneuve, L.M.; Morris, K.V.; Rossi, J.J. Argonaute-1 directs siRNA-mediated transcriptional gene silencing in human cells. *Nat. Struct. Mol. Biol.* **2006**, *13*, 793–797. [CrossRef] [PubMed]

37. Younger, S.T.; Corey, D.R. Transcriptional gene silencing in mammalian cells by miRNA mimics that target gene promoters. *Nucleic Acids Res.* **2011**, *39*, 5682–5691. [CrossRef]

38. Zardo, G.; Ciolfi, A.; Vian, L.; Starnes, L.M.; Billi, M.; Racanicchi, S.; Maresca, C.; Fazi, F.; Travaglini, L.; Noguera, N.; et al. Polycombs and microRNA-223 regulate human granulopoiesis by transcriptional control of target gene expression. *Blood* **2012**, *119*, 4034–4046. [CrossRef]

39. Huang, V.; Zheng, J.; Qi, Z.; Wang, J.; Place, R.F.; Yu, J.; Li, H.; Li, L.C. Ago1 Interacts with RNA Polymerase II and Binds to the Promoters of Actively Transcribed Genes in Human Cancer Cells. *PLoS Genet.* **2013**, *9*, e1003821. [CrossRef]

40. Miao, L.; Yao, H.; Li, C.; Pu, M.; Yao, X.; Yang, H.; Qi, X.; Ren, J.; Wang, Y. A dual inhibition: MicroRNA-552 suppresses both transcription and translation of cytochrome P450 2E1. *Biochim. Biophys. Acta-Gene Regul. Mech.* **2016**, *1859*, 650–662. [CrossRef]

41. Cora, D.; Re, A.; Caselle, M.; Bussolino, F. MicroRNA-mediated regulatory circuits: Outlook and perspectives. *Phys. Biol.* **2017**, *14*, 045001. [CrossRef] [PubMed]

42. Hashim, A.; Rizzo, F.; Marchese, G.; Ravo, M.; Tarallo, R.; Nassa, G.; Giurato, G.; Santamaria, G.; Cordella, A.; Cantarella, C.; et al. RNA sequencing identifies specific PIWI-interacting small noncoding RNA expression patterns in breast cancer. *Oncotarget* **2014**, *5*, 9901–9910. [CrossRef] [PubMed]

43. Alexandrova, E.; Miglino, N.; Hashim, A.; Nassa, G.; Stellato, C.; Tamm, M.; Baty, F.; Brutsche, M.; Weisz, A.; Borger, P. Small RNA profiling reveals deregulated phosphatase and tensin homolog (PTEN)/phosphoinositide 3-kinase (PI3K)/Akt pathway in bronchial smooth muscle cells from asthmatic patients. *J. Allergy Clin. Immunol.* **2016**, *137*, 58–67. [CrossRef] [PubMed]

44. Ross, R.J.; Weiner, M.M.; Lin, H. PIWI proteins and PIWI-interacting RNAs in the soma. *Nature* **2014**, *505*, 353–359. [CrossRef] [PubMed]

45. Moore, R.S.; Kaletsky, R.; Murphy, C.T. Piwi/PRG-1 argonaute and TGF-β mediate transgenerational learned pathogenic avoidance. *Cell* **2019**, *177*, 1827–1841. [CrossRef]

46. Posner, R.; Toker, I.A.; Antonova, O.; Star, E.; Anava, S.; Azmon, E.; Hendricks, M.; Bracha, S.; Gingold, H.; Rechavi, O. Neuronal Small RNAs Control Behavior Transgenerationally. *Cell* **2019**, *177*, 1814–1826. [CrossRef]

Review

Features of DNA Repair in the Early Stages of Mammalian Embryonic Development

Evgenia V. Khokhlova [1,2], Zoia S. Fesenko [1], Julia V. Sopova [1,3] and Elena I. Leonova [1,4,*]

[1] Institute of Translational Biomedicine, St. Petersburg State University, 199034 St. Petersburg, Russia; evkhokhlova95@gmail.com (E.V.K.); zozoya07@mail.ru (Z.S.F.); sopova@hotmail.com (J.V.S.)

[2] Institute of Cytology of the Russian Academy of Sciences, 194064 St. Petersburg, Russia

[3] Laboratory of Amyloid Biology, St. Petersburg State University, 199034 St. Petersburg, Russia

[4] Preclinical Research Center, University of Science and Technology, 1 Olympic Ave, 354340 Sochi, Russia

* Correspondence: e.leonova@spbu.ru; Tel.: +8-(999)-232-92-58

Received: 29 August 2020; Accepted: 25 September 2020; Published: 27 September 2020

Abstract: Cell repair machinery is responsible for protecting the genome from endogenous and exogenous effects that induce DNA damage. Mutations that occur in somatic cells lead to dysfunction in certain tissues or organs, while a violation of genomic integrity during the embryonic period often leads to death. A mammalian embryo's ability to respond to damaged DNA and repair it, as well as its sensitivity to specific lesions, is still not well understood. In this review, we combine disparate data on repair processes in the early stages of preimplantation development in mammalian embryos.

Keywords: DNA repair; BER (base excision repair); NER (nucleotide excision repair); MMR (mismatch repair); DSBR (double strand break repair); HR (homologous recombination); NHEJ (nonhomologous end joining); MHEJ (microhomologies end joining); oocyte; zygote; blastocyst

1. Introduction

DNA repair during the early stages of embryonic development has one of the most significant effects on embryonic fate [1]. In the early embryonic stages of development, cells differ in their threshold of sensitivity to endogenous and exogenous factors [2]. However, to preserve and maintain the integrity of the genome, cells activate complex DNA repair mechanisms. It is believed that all major DNA repair pathways function in embryos. Repair proteins interact with cell cycle control proteins to stop the cell cycle if DNA is damaged, allowing the repair complexes time to fix the damaged DNA. If the damage is too substantial, and it is impossible to repair the DNA, a proapoptotic pathway is activated, resulting in cell death [3]. The mechanisms of regulation and functioning of repair systems are well studied in somatic cells. However, less is known about their activities during early embryonic development. Several reports have shown that early embryos and embryonic stem (ES) cells lack functional cell cycle control checkpoints, and DNA synthesis and cell division continue in the presence of damaged DNA. Ineffective activation of cell cycle checkpoints and suppression of apoptotic pathways in early embryos is associated with a shortened cell cycle, helping to ensure that the first embryonic cell division occurs, even under adverse conditions [4]. Thus, this review aims to analyze the literature and compare the role of repair systems at different stages of early mammalian embryonic development from the oocyte to the preimplantation blastocyst.

2. Oocyte Repair

Oocytes are one of the longest-living cells in the body, remaining at rest for many months (mouse) or decades (humans) [5]. During this time, they are exposed to exogenous and endogenous factors that cause damage to the DNA structure. DNA double strand breaks (DSBs) accumulate with age in primary follicle oocytes due to cellular metabolism and oxidative stress [6]. Factors such as γ-radiation, chemotherapy, and adverse environmental influences lead to the formation of DSBs in oocytes during the primary follicular stage [7–9]. In each of these cases, DSBs induce oocyte death if the damaged DNA is not repaired. This leads to depletion of the oocyte pool in the follicles, premature ovarian failure, infertility, and early menopause. Primary follicular oocytes at the germinal vesicle (GV) stage are more susceptible to DNA-damaging agents and are more prone to apoptosis compared to somatic cells and more mature MII stage oocytes [10]. This might be associated with the development of a highly sensitive apoptotic response since it is crucial to eliminate oocytes with damaged DNA to protect the germline [11]. Therefore, DNA damage control checkpoints are activated when the cell cycle stops during meiosis I, facilitating removal of oocytes with DNA that has not been restored after meiotic recombination. Members of the p53 family have been identified as critical regulators of apoptosis activity in oocytes during the GV stage [12]. According to published data, the *TAp63* gene is highly expressed in primary follicle oocytes and is an essential mediator of induced DNA damage response in oocytes due to transcriptional activation of proapoptotic members of the Bcl-2 family, PUMA and NOXA [13]. Interestingly, *TAp63* expression is suppressed when oocytes exit the follicle, which may partially explain why mature MII oocytes are more resistant to apoptosis due to DNA damage than oocytes in the GV stage (Figure 1) [14].

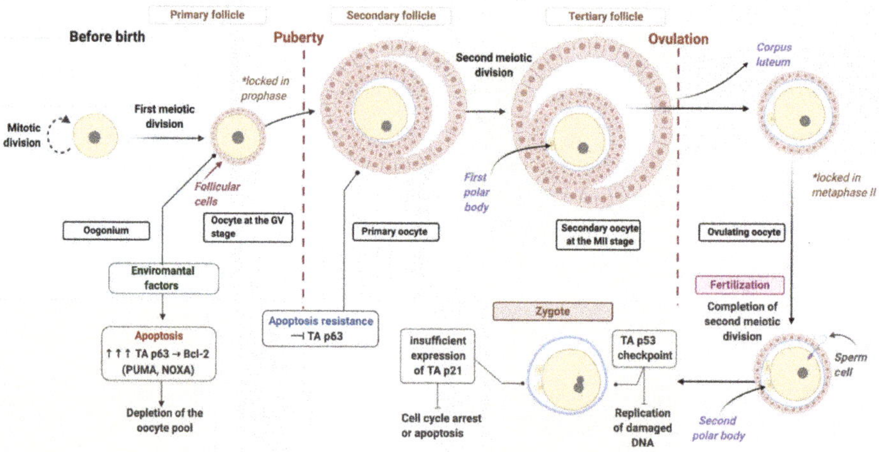

Figure 1. The process of differentiation in female germ cells. Created with BioRender.com.

Furthermore, mature MII oocytes have a broader expression profile of mRNA encoding repair proteins compared to GV oocytes. Nevertheless, starting from the oocyte at the GV stage, expression of genes encoding proteins for all repair systems has been observed [15]. Extensive expression of repair genes corresponds to the oocyte's ability to recognize and repair DNA damage from the earliest stages of development [16]. In 2009, Jaroudi et al. demonstrated that in humans, mRNA levels of most repair genes in oocytes are higher than in blastocysts, which is explained by the accumulation of a sufficient amount of mRNA to ensure preservation of the genome before and after fertilization until the zygotic genome is activated [17]. DNA repair transcripts that accumulate in human oocytes play an essential role in chromatin remodeling and maintenance of chromatin integrity during fertilization [18]. Transcripts of all DNA repair pathways, including base excision repair (BER), mismatch repair (MMR), nucleotide excision repair (NER), double strand break (DSBs) repair are presented in oocytes at the GV and MII stages in mouse, monkey, and human [17,19,20]. DNA glycosylase is involved in base excision repair in oocytes and zygotes and exhibits significantly higher levels in MII oocytes than in oocytes at earlier stages [21]. Similarly, expression levels of the *XPC* gene involved in nucleotide excision repair in MII oocytes is significantly higher than in GV oocytes [22]. Expression of the mismatch repair gene *MSH2* was also higher in oocytes of stage MII compared with oocytes in stages GV and MI [21]. *ATM* and *ATR* DNA repair markers are actively expressed during oocyte maturation, *ATR* expression is primarily manifested in immature oocytes during meiosis I. In addition, the DNA repair marker *BARD1* is highly expressed during oocyte maturation [23]. The BARD1-BRCA1 heterodimer is considered as an E3 ubiquitin ligase. Studies by Gasca et al. found that the E3 ubiquitin ligase multiprotein complex containing BARD1, BRCA1, and BRCA2 is involved in the maturation of human oocytes [24]. These proteins play a crucial role in regulating cell cycle development, DNA repair, and gene transcription [25]. Mutations in *BRCA* genes lead to an impaired ability to repair DNA DSBs and cause premature oocyte aging, apoptosis, and disturbances in meiotic divisions [26]. *BRCA1* and *BRCA2* homozygous deletion in mice results in embryonic lethality. Heterozygous deletion of *BRCA1* in mice results in impaired reproductive capacity, characterized by low follicle counts and an increase in the accumulation of DNA DSBs in surviving follicles relative to wild-type mice [10]. In experiments on rhesus monkeys, the RAD51 protein, which is involved in homologous DSB recombination, is expressed in oocytes. However, its expression decreases during oocyte maturation and then increases again at the eight-cell stage [21]. *RAD51* homozygous deletion in mice results in defective separation of sister chromatids, aneuploidy, and broken chromosomes at metaphase II [10].

According to Jaroudi et al., *RAD51* and *MSH2* are expressed at high levels in both oocytes and human blastocysts. The Ku70 is the DNA-binding component of the non-homologous end joining (NHEJ) repair machinery, it exhibits high expression levels in both oocytes and blastocysts [17]. Studying human oocytes using the single-cell sequencing method, the *DPYD* gene was discovered, which encodes the dihydropyrimidine dehydrogenase. Importantly, high NADP+ levels activated *DPYD* to enhance the repair of DNA double-strand breaks to maintain euploidy. In vivo high expression level of *DPYD* was observed in primary and secondary follicle oocytes, but this gene was not expressed in oocytes and preimplantation embryos. Expression of *DPYD* increased in primary and secondary follicles incubated in vitro and was dramatically upregulated in *in vitro* matured (IVM) oocytes. Furthermore, embryos from human IVM oocytes had more tiny chromosomal defects than those from human *in vivo* matured (IVO) oocytes. It has been shown that increasing the expression of *DPYD* may facilitate the repair process and overcome the risk of aneuploidy [27]. The compounds of *in vitro* culture medium may have an impact on embryo's viability by dysregulation of some genes associated with DNA repair machinery [28]. Although all repair pathways function in oocytes, each pathway's activation may vary depending on the stage of oocyte development [10]. Additionally, some reactions to DNA damage and DNA repair genes are overrepresented in the oocyte compared to the preimplantation embryo (from one-cell to blastocyst stages) [28]. This might reflect the particular importance of ensuring the integrity of the genome, especially after prolonged arrest during the first meiotic prophase. The repair pathways are redundant, and it is unknown whether all pathways are used simultaneously during oocyte development. Some mRNAs are translated in the oocyte, while others remain in a polyadenylated form until fertilization, maintaining a pool of repair proteins until the embryo genome is activated [29].

3. Repair at the Zygote Stage

The transformation of a fertilized oocyte into a zygote is an amazing process that occurs in the absence of transcription and depends on the mRNA accumulated in the oocyte during oogenesis. Sperm penetration causes activation of the egg, accompanied by the release of mRNA from complexes that block translation initiation [30]. In mouse oocytes, mRNAs contain elements of cytoplasmic polyadenylation (CPE) in their 3′-untranslated region [31]. Polyadenylated mRNA tails are linked to a repression translation complex containing CPE-binding protein (CPEB) and Maskin protein. Maskin binds eukaryotic translation initiation factor 4E (eIF4E), an interaction that excludes eIF4G and prevents formation of the eIF4F initiation complex [32]. During oocyte maturation, CPEB phosphorylation stimulates polyadenylation and recruitment of the poly (A)-binding protein bound to eIF4G, which helps to displace Maskin from eIF4E, thereby initiating translation [32]. The activation process contributes to the continuation of meiosis, the formation of pronuclei, and the translation of proteins necessary for the zygote [30]. Thus, the fusion of a terminally differentiated oocyte in metaphase II and the sperm cause complex changes, including chromatin remodeling and epigenetic reprogramming in the zygote. The most significant changes occur in the paternal genome, where compacted sperm chromatin is reorganized, protamines are replaced by histones, and various histones and DNA are demethylated in phases G1 and S [33]. At this stage in the chromosomes separately in the male and female pronuclei, active demethylation of the paternal DNA and passive demethylation of the maternal DNA occur [34]. Active demethylation of the paternal genome's DNA includes mechanisms based on excision and DNA repair [35]. It should be noticed that cytosines in sperm DNA are highly methylated. Moreover, the majority of 5mC is demethylated, regardless of DNA replication, before the first cycle of zygotic cell division [36]. The mechanisms of active DNA demethylation in zygotes are not well understood. It has been suggested that different DNA-repair-based mechanisms are important for demethylation, e.g., processes initiated by DNA glycosylases, DNA methyl-transferases and DNA deaminases [37]. During zygotic reprogramming, 5mC is modified to 5-hydroxymethylcytosine (5hmC) using Tet3 hydroxylase. This oxidized cytosine is either removed or replaced with unmodified cytosine using BER or is passively removed during the next DNA replication cycle [38]. This leads to the breaking of DNA chains. Double- and single-stranded DNA breaks lead to the appearance of phosphorylated histone H2AX (γH2AX) and activation of other DNA repair systems [39]. Mature sperm cells are considered incapable of repairing DNA damage due to DNA compaction and decreased transcriptional activity. It has been suggested that the oocyte has the capacity to repair sperm DNA damage when the level of sperm DNA damage is less than 8%. Higher levels of sperm DNA damage are associated with a failure to reach the blastocysts phase and embryonic loss between the embryonic genome activation (EGA) and the blastocyst stages. [40]. There is evidence that human sperm has a basic BER pathway containing only OGG1 protein [41]. Being the first enzyme in the pathway, its presence is sufficient for sperm to detect and remove oxidized bases and residues of 8-OHdG, a standard oxidative stress product. Because the rest of the pathway is truncated, DNA obtained after removal of 8-OHdG must subsequently be restored by oocyte repair proteins after fertilization before the first cell division. However, it is interesting to note that expression of the *OGG1* gene in the oocyte at this stage is low [42]. Some authors consider the random complementarity of sperm cells and oocytes to be an elaborate mechanism for checking the compatibility of oocytes and fertilizing sperm since both must be involved in the restoration of oxidative damage to DNA [40]. It is believed that BER is the primary pathway for the restoration of oxidative damage in the zygotic genome. However, other repair systems can also contribute to the restoration of embryonic DNA at this stage of development [6]. DSB restoration in the zygote occurs using NHEJ and homologous recombination (HR) repair pathways. These pathways are not equally important during the cell cycle. The choice of DSB repair pathway depends on the developmental stage of the embryo and the cell cycle. NHEJ works throughout the cell cycle, while HR functions during the S/G2 stage. DSBs obtained by stopping replication are preferably restored using HR [43]. It is believed that at the zygotic stage, NHEJ plays an essential role in the restoration of sperm DSBs [44].

During fertilization male and female genomes fuse to form a zygote nucleus. Dynamic chromatin and protein rearrangements require post-translational modification, such as poly(ADP-ribosyl)ation, for the post-fertilization development. In addition, poly(ADP-ribosyl)ation of nuclear proteins is necessary for the detection of DNA strand breaks and the recruitment of repair factors to damaged sites. Poly(ADP-ribosyl)ation is catalyzed by enzymes of the poly(ADP-ribose) polymerase-1 (PARP) family. Cellular stress stimulates the activity of poly(ADP-ribose) polymerase-1 (PARP-1) on binding to DNA strand breaks, playing a key role in their repair. Inhibition of PARP-1 in oocytes impacts negatively on embryo survival [45]. Mice with homozygous mutations in the *PARP-1* gene are viable, although cells lacking *PARP-1* are hypersensitive to many DNA-damaging agents, and double homozygous knockout of *PARP-1* and *PARP-2* leads to embryonic mortality [45]. PARP-1 is one of the first proteins involved in the repair of DSBs based on micro-homologous end joining (MHEJ). PARP-1 has an affinity for the 3′-ends of DNA and competes with Ku proteins for binding to DNA upon damage, switching DSB repair pathways from NHEJ to MHEJ [46,47].

Normally, the paternal and maternal pronuclei in the mouse zygotes initiate DNA replication nearly synchronously between five and six hours after fertilization. But in response to DNA damage in the male pronucleus the zygote reacts by slowing down the replication of paternal DNA for up to 12 h, ultimately leading to a halt in embryonic development. Moreover, the replication delay in the female pronucleus does not occur. Pronuclei act independently, stopping DNA synthesis in response to DNA damage in only one pronucleus. The lack of synchronization in DNA replication between the two pronuclei in response to damage in the paternal DNA leads to two pronuclei at different stages of DNA replication being in the same cell [48]. Low expression of *CDKN1A* (p21) gene at the early embryo stage may result in inability of the embryo to respond to DNA damage by stopping the cell cycle. p21 is a target of p53 and mediates p53-dependent cell cycle arrest or apoptosis. The p53-dependent S-phase checkpoint functions at the zygotic stage to inhibit replication of damaged DNA [16]. p21-mediated cell cycle arrest occurs later, preventing delayed chromosome damage. Thus, during early development, embryos are protected by mechanisms regulated by p53 and p21 [49]. The first signs of apoptosis, such as cytoplasmic fragmentation, in the case of excessive unrepaired DNA damage, appear only at the two-cell stage in mice and four-cell stage in humans. However, other signs of apoptosis, including condensation of chromatin and cytoplasm with subsequent DNA degradation and nuclear fragmentation, are not observed until the morula and blastocyst stages (Figure 1) [48].

4. Repair at the Cleavage and Blastocyst Stages

Human embryos at cleavage state show a high level of postzygotic chromosomal mosaicism, including aneuploidy and polyploidy. Mosaicism for the paternal alleles most probably resulted from mutations in genes that are involved in DNA mismatch repair (MMR), e.g., *PMS2* and *MSH2* [50,51]. The gene expression profiles of oocytes and one-cell embryos are highly similar, at the two-cell stage the gene expression dramatically changes. The second change in gene expression of embryonic genomes occurs at the four to eight cell stage, preceding cell compaction at the morula stage, which explains the separation between the two-cell and eight-cell stages. Eight-cell embryos and blastocysts slightly differ in gene expression (Figure 2) [28].

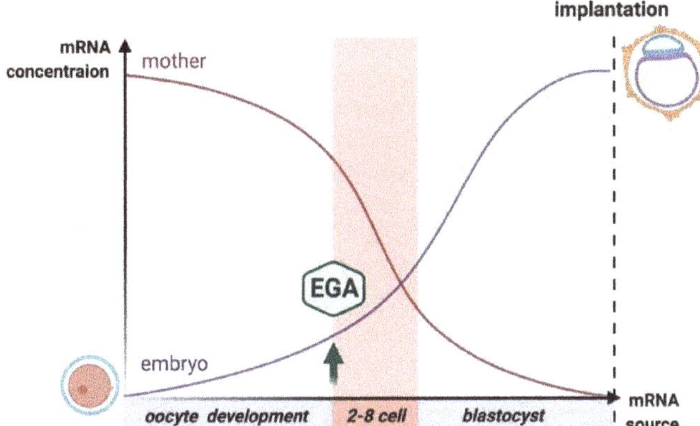

Figure 2. The source and dynamics of mRNA expression during early embryonic development. EGA—embryonic genome activation. Created with BioRender.com.

Blastocyst formation is the first stage during which the differentiation of cells into two types occurs, characterized by differences in gene expression between inner cell mass (ICM) and trophectoderm cells (TE) cells [28]. Subsequently, TE forms tissues of the placenta, while ICM cells give rise to embryonic tissues. Since ICM cells give rise to cells forming a new organism, maintaining the integrity of the genome in these cells is crucial [1]. Due to the high replication rate and the beginning of cell differentiation in the blastocyst, the expression profile of DNA repair genes differs from oocytes. It is believed that all DNA repair pathways are present in both the human oocyte and in the blastocyst. However, some DNA repair genes are expressed in the blastocyst at lower levels compared to the oocyte, such as *MBD4, NEIL1, OGG1* (BER); *RAD50, RAD54B, RBBP8* (HR); *MSH3* (MMR); *ERCC5, GTF2H2, LIG1, RPA1* (NER) [17]. This may be explained by the fact that the oocyte contains maternal mRNA transcripts to maintain genome integrity before EGA [17]. During development, embryos acquire the ability to respond to factors that cause DNA damage by activating and regulating DNA repair and apoptosis genes, similar to what occurs in somatic cells [16]. Mutations in some DNA repair genes lead to defects in early and later stages, including fetal mortality, infertility, or cancer susceptibility in postembryonic development. However, the fact that not all mutations in repair genes delay embryonic development in the early stages of development indicates that some mechanisms of DNA repair are either redundant or play less active roles during early embryogenesis and rather, may be involved during later developmental stages. Detailed information about the primary patterns of expression of DNA repair genes during different stages of development is needed to understand the extent to which preimplantation embryos can react to and repair DNA damage and to what extent preimplantation embryos are selectively sensitive to certain forms of DNA damage [16]. This information is difficult to obtain because the concentration of mRNA expressed in early embryos is quite low. However, many protocols have recently been developed, for instance, single-cell sequencing, that allow detection of small amounts of mRNA in each individual cell. Recently, many articles using this method have appeared [52–54]. Another way to investigate repair pathways in the early stages of development is to investigate blastocyst-derived embryonic stem (ES) cells.

5. Embryonic Stem Cells

ES cells are pluripotent cells isolated from the inner cell mass of a blastocyst. ES cells are sensitive to DNA damage and easily undergo apoptosis, removing damaged cells from the pluripotent pool [55]. Additional evidence suggests that DNA damage can cause premature differentiation in these cells. ES cells have a robust set of DNA repair mechanisms [20]. In particular, ES cells maintain significantly higher expression levels of proteins associated with HR than their expression levels in differentiated cells. Additionally, it is believed that levels of HR proteins decrease as ES cells differentiate [56]. ES cells have very short G1 phase and long period of S phase time can promote the use of HR rather than NHEJ since many of the proteins involved in HR also participate in DNA replication [57]. Due to the short G1 phase, sister chromatids are available for efficient recombination-mediated repair in ES cells [55]. ES cells derived from mouse blastocysts (mES) synthesize constantly high level of HR proteins throughout the cell cycle [56]. It can be suggested that the active HR repair pathway is required for a rapid response to DNA breaks in ES cells [56]. HR proteins are thought to be highly expressed but remain inactive until posttranslational modifications are triggered in response to damage. In experiments conducting treatment of mES cells with DNA break-inducing agents, Rad51, Rad54, Exo1, and γH2AX proteins were redistributed and concentrated in the nucleus as discrete foci. Therefore, in response to DNA damage, Rad51, Rad54, and Exo1 proteins are immediately localized at the sites of DNA strand break regions. It is believed that Rad51, Rad54, and Exo1 are constantly present in mES cells, providing fast and efficient HR-mediated DNA repair when needed [56]. The MMR also seems to play an essential role in ES cells, determining the fate of the cell with respect to whether the cell follows the DNA repair pathway or undergoes apoptosis. High endogenous levels of MSH2 protein in ES cells promote apoptosis, while low levels promote DNA repair, as occurs in differentiated cells [55]. Even low levels of MSH2 protein can reduce the number of spontaneous mutations in ES cells compared to MSH2 knockout ES cells, suggesting that MMR plays a fundamental role in regulating the level of mutagenesis in preimplantation embryonic cells [55]. NER machinery of ES cells cannot repair damage after high UV doses resulting in a rapid induction of apoptosis [58]. It is also known that the G1/S

DNA damage checkpoint in ES cells can be activated upon DNA damage and prevent the cell from passing into the S phase [56]. However, there are conflicting data on this subject. Other studies indicate that rapidly dividing embryos and mouse ES cells have a nonfunctional G1/S checkpoint. Therefore, they cannot stop at G1 in the presence of DNA damage [59]. It is believed that high levels of the protein phosphatase CDC25A and the downstream CHK1 effector modulate the efficacy of the G1/S checkpoint in murine ES cells [60]. The ubiquitin hydrolase DUB3/USP17L2 supports high levels of CDC25A in these cells by removing polyubiquitin chains. The *DUB3* gene itself is a target for two pluripotency factors, ESRRβ and SOX2, and its expression is regulated during development, followed by rapid suppression after cell differentiation [61]. Therefore, the nonfunctioning control point G1/S is characteristic of mouse ES cells that are inappropriately associated with the state of pluripotency [62]. However, in human ES cells, the G1/S checkpoint is active, while the S-phase checkpoint is inactive. This difference may be due to differences in mouse and human embryo cell cycle and differences in pluripotency maintenance. It may also be the reason for differences in the activation of the embryonic genome. In mouse, EGA occurs at the two-cell stage, while in human EGA occurs at the eight-cell stage. Following this hypothesis, it was recently reported that primates have less reliable mechanisms for genome surveillance than rodents [59].

6. Conclusions

In this review, we briefly survey the repair processes from the oocyte stage to the preimplantation stage of the blastocyst. In each of these stages, we see changes in gene expression patterns and the involved repair systems. Oocytes in the GV stage are more susceptible to DNA damage and more prone to apoptosis, while more mature stage II oocytes do not activate apoptosis and are more resistant to DNA damage [10]. At MII stage, oocytes actively accumulate transcripts of repair proteins to protect the genome from DNA damage during fertilization and the first division. After fertilization in zygotes, maternal mRNAs that accumulated in the oocyte are actively used to repair the sperm genome and to maintain DNA integrity in subsequent development. Antiapoptotic proteins that protect the early embryo from death are also active at this stage because apoptosis at this stage would be fatal. However, in cases of critical levels of damage, developmental arrest occurs, accompanied by cell cycle arrest [48]. As the number of embryonic cells increases, the embryonic genome is activated, and maternal mRNAs are destroyed [18]. During this period, the embryo becomes more sensitive to external influences. However, when passing from the stage of a two-cell embryo to the blastocyst stage, its own repair proteins accumulate, and apoptotic systems that remove blastomeres with damage are activated [63].

As cells differentiate and lose their pluripotency, they acquire repair properties similar to somatic cells. An active change in the patterns of expression of repair genes during embryonic development indicates that repair processes are essential for normal functions at all stages of development. Repair processes in embryonic development are especially important. Currently, there are many gaps regarding the precise roles and timing of expression of some DNA repair genes in the early stages of embryonic development. The observed stage-specific variations in transcripts and expressed proteins of DNA repair genes indicate the difficulty in regulating these pathways during development.

Author Contributions: Lead the project, contributed to the conceptualization, supervision and funding acquisition for the project, E.I.L.; processed the data and wrote the review, E.V.K. and J.V.S.; processed the data and produced the figures Z.S.F. All authors read and approved the final version of the manuscript.

Funding: The reported study was funded by RFBR, project number 19-315-51030; This work was performed within project ID: 60810083 of the St. Petersburg State University, St. Petersburg, Russia.

Conflicts of Interest: All authors declare no conflict of interest.

References

1. Juan, H.-C.; Lin, Y.; Chen, H.-R.; Fann, M.-J. Cdk12 is essential for embryonic development and the maintenance of genomic stability. *Cell Death Differ.* **2016**, *23*, 1038–1048. [CrossRef] [PubMed]
2. Hamdoun, A.; Epel, D. Embryo stability and vulnerability in an always changing world. *Proc. Natl. Acad. Sci. USA* **2007**, *104*, 1745–1750. [CrossRef]
3. Smith, M.L.; Fornace, A.J. Mammalian DNA damage-inducible genes associated with growth arrest and apoptosis. *Mutat. Res. Rev. Genet. Toxicol.* **1996**, *340*, 109–124. [CrossRef]
4. Harrison, R.H.; Kuo, H.-C.; Scriven, P.N.; Handyside, A.H.; Ogilvie, C.M. Lack of cell cycle checkpoints in human cleavage stage embryos revealed by a clonal pattern of chromosomal mosaicism analysed by sequential multicolour FISH. *Zygote* **2000**, *8*, 217–224. [CrossRef] [PubMed]
5. Rimon-Dahari, N.; Yerushalmi-Heinemann, L.; Alyagor, L.; Dekel, N. Ovarian Folliculogenesis: Molecular Mechanisms of Cell Differentiation in Gonad Development. In *Results and Problems in Cell Differentiation*; Piprek, R.P., Ed.; Springer International Publishing: Cham, Switzerland, 2016; Volume 58, pp. 167–190, ISBN 978-3-319-31971-1.
6. Martin, J.H.; Bromfield, E.G.; Aitken, R.J.; Nixon, B. Biochemical alterations in the oocyte in support of early embryonic development. *Cell. Mol. Life Sci.* **2017**, *74*, 469–485. [CrossRef]
7. Kerr, J.B.; Brogan, L.; Myers, M.; Hutt, K.J.; Mladenovska, T.; Ricardo, S.; Hamza, K.; Scott, C.L.; Strasser, A.; Findlay, J.K. The primordial follicle reserve is not renewed after chemical or γ-irradiation mediated depletion. *Reproduction* **2012**, *143*, 469–476. [CrossRef] [PubMed]
8. Kujjo, L.L.; Laine, T.; Pereira, R.J.G.; Kagawa, W.; Kurumizaka, H.; Yokoyama, S.; Perez, G.I. Enhancing Survival of Mouse Oocytes Following Chemotherapy or Aging by Targeting Bax and Rad51. *PLoS ONE* **2010**, *5*, e9204. [CrossRef]
9. Hunt, P.A.; Lawson, C.; Gieske, M.; Murdoch, B.; Smith, H.; Marre, A.; Hassold, T.; VandeVoort, C.A. Bisphenol A alters early oogenesis and follicle formation in the fetal ovary of the rhesus monkey. *Proc. Natl. Acad. Sci. USA* **2012**, *109*, 17525–17530. [CrossRef]
10. Winship, A.L.; Stringer, J.M.; Liew, S.H.; Hutt, K.J. The importance of DNA repair for maintaining oocyte quality in response to anti-cancer treatments, environmental toxins and maternal ageing. *Hum. Reprod. Update* **2018**, *24*, 119–134. [CrossRef]
11. Stringer, J.M.; Winship, A.; Liew, S.H.; Hutt, K. The capacity of oocytes for DNA repair. *Cell. Mol. Life Sci.* **2018**, *75*, 2777–2792. [CrossRef]
12. Hu, W. The Role of p53 Gene Family in Reproduction. *Cold Spring Harb. Perspect. Biol.* **2009**, *1*, a001073. [CrossRef] [PubMed]
13. Kerr, J.B.; Hutt, K.J.; Michalak, E.M.; Cook, M.; Vandenberg, C.J.; Liew, S.H.; Bouillet, P.; Mills, A.; Scott, C.L.; Findlay, J.K.; et al. DNA Damage-Induced Primordial Follicle Oocyte Apoptosis and Loss of Fertility Require TAp63-Mediated Induction of Puma and Noxa. *Mol. Cell* **2012**, *48*, 343–352. [CrossRef] [PubMed]
14. Suh, E.-K.; Yang, A.; Kettenbach, A.; Bamberger, C.; Michaelis, A.H.; Zhu, Z.; Elvin, J.A.; Bronson, R.T.; Crum, C.P.; McKeon, F. p63 protects the female germ line during meiotic arrest. *Nature* **2006**, *444*, 624–628. [CrossRef] [PubMed]
15. Wang, S.; Kou, Z.; Jing, Z.; Zhang, Y.; Guo, X.; Dong, M.; Wilmut, I.; Gao, S. Proteome of mouse oocytes at different developmental stages. *Proc. Natl. Acad. Sci. USA* **2010**, *107*, 17639–17644. [CrossRef]
16. Zheng, P.; Schramm, R.D.; Latham, K.E. Developmental Regulation and In Vitro Culture Effects on Expression of DNA Repair and Cell Cycle Checkpoint Control Genes in Rhesus Monkey Oocytes and Embryos. *Biol. Reprod.* **2005**, *72*, 1359–1369. [CrossRef] [PubMed]
17. Jaroudi, S.; Kakourou, G.; Cawood, S.; Doshi, A.; Ranieri, D.M.; Serhal, P.; Harper, J.C.; SenGupta, S.B. Expression profiling of DNA repair genes in human oocytes and blastocysts using microarrays. *Hum. Reprod.* **2009**, *24*, 2649–2655. [CrossRef]
18. Tulay, P.; Naja, R.P.; Cascales-Roman, O.; Doshi, A.; Serhal, P.; SenGupta, S.B. Investigation of microRNA expression and DNA repair gene transcripts in human oocytes and blastocysts. *J. Assist. Reprod. Genet.* **2015**, *32*, 1757–1764. [CrossRef]
19. Menezo, Y.J.; Russo, G.; Tosti, E.; Mouatassim, S.E.; Benkhalifa, M. Expression profile of genes coding for DNA repair in human oocytes using pangenomic microarrays, with a special focus on ROS linked decays. *J. Assist. Reprod. Genet.* **2007**, *24*, 513–520. [CrossRef]

20. Zeng, F.; Baldwin, D.A.; Schultz, R.M. Transcript profiling during preimplantation mouse development. *Dev. Biol.* **2004**, *272*, 483–496. [CrossRef]

21. Men, N.T.; Kikuchi, K.; Furusawa, T.; Dang-Nguyen, T.Q.; Nakai, M.; Fukuda, A.; Noguchi, J.; Kaneko, H.; Viet Linh, N.; Xuan Nguyen, B.; et al. Expression of DNA repair genes in porcine oocytes before and after fertilization by ICSI using freeze-dried sperm: DNA Repair Genes in Porcine Oocytes. *Anim. Sci. J.* **2016**, *87*, 1325–1333. [CrossRef]

22. Gunes, S.; Sertyel, S. Sperm DNA Damage and Oocyte Repair Capability. In *A Clinician's Guide to Sperm DNA and Chromatin Damage*; Zini, A., Agarwal, A., Eds.; Springer International Publishing: Cham, Switzerland, 2018; pp. 321–346. ISBN 978-3-319-71814-9.

23. Virant-Klun, I.; Knez, K.; Tomazevic, T.; Skutella, T. Gene Expression Profiling of Human Oocytes Developed and Matured In Vivo or In Vitro. *Biomed Res. Int.* **2013**, *2013*, 1–20. [CrossRef]

24. Gasca, S.; Pellestor, F.; Assou, S.; Loup, V.; Anahory, T.; Dechaud, H.; De Vos, J.; Hamamah, S. Identifying new human oocyte marker genes: A microarray approach. *Reprod. Biomed. Online* **2007**, *14*, 175–183. [CrossRef]

25. Lee, N.S.; Kim, S.; Jung, Y.W.; Kim, H. Eukaryotic DNA damage responses: Homologous recombination factors and ubiquitin modification. *Mutat. Res. Fundam. Mol. Mech. Mutagenesis* **2018**, *809*, 88–98. [CrossRef] [PubMed]

26. Giritharan, G.; Talbi, S.; Donjacour, A.; Di Sebastiano, F.; Dobson, A.T.; Rinaudo, P.F. Effect of in vitro fertilization on gene expression and development of mouse preimplantation embryos. *Reproduction* **2007**, *134*, 63–72. [CrossRef]

27. Zhao, H.; Li, T.; Zhao, Y.; Tan, T.; Liu, C.; Liu, Y.; Chang, L.; Huang, N.; Li, C.; Fan, Y.; et al. Single-Cell Transcriptomics of Human Oocytes: Environment-Driven Metabolic Competition and Compensatory Mechanisms During Oocyte Maturation. *Antioxid. Redox Signal.* **2019**, *30*, 542–559. [CrossRef]

28. Jaroudi, S.; SenGupta, S. DNA repair in mammalian embryos. *Mutat. Res. Rev. Mutat. Res.* **2007**, *635*, 53–77. [CrossRef]

29. Ménézo, Y.; Dale, B.; Cohen, M. DNA damage and repair in human oocytes and embryos: A review. *Zygote* **2010**, *18*, 357–365. [CrossRef]

30. Stitzel, M.L.; Seydoux, G. Regulation of the Oocyte-to-Zygote Transition. *Science* **2007**, *316*, 407–408. [CrossRef]

31. Dai, X.-X.; Jiang, J.-C.; Sha, Q.-Q.; Jiang, Y.; Ou, X.-H.; Fan, H.-Y. A combinatorial code for mRNA 3′-UTR-mediated translational control in the mouse oocyte. *Nucleic Acids Res.* **2019**, *47*, 328–340. [CrossRef]

32. Stebbins-Boaz, B.; Cao, Q.; de Moor, C.H.; Mendez, R.; Richter, J.D. Maskin Is a CPEB-Associated Factor That Transiently Interacts with eIF-4E. *Mol. Cell* **1999**, *4*, 1017–1027. [CrossRef]

33. Bao, J.; Bedford, M.T. Epigenetic regulation of the histone-to-protamine transition during spermiogenesis. *Reproduction* **2016**, *151*, R55–R70. [CrossRef] [PubMed]

34. Guo, F.; Li, X.; Liang, D.; Li, T.; Zhu, P.; Guo, H.; Wu, X.; Wen, L.; Gu, T.-P.; Hu, B.; et al. Active and Passive Demethylation of Male and Female Pronuclear DNA in the Mammalian Zygote. *Cell Stem Cell* **2014**, *15*, 447–459. [CrossRef] [PubMed]

35. Zhu, J.-K. Active DNA Demethylation Mediated by DNA Glycosylases. *Annu. Rev. Genet.* **2009**, *43*, 143–166. [CrossRef] [PubMed]

36. Wossidlo, M.; Arand, J.; Sebastiano, V.; Lepikhov, K.; Boiani, M.; Reinhardt, R.; Schöler, H.; Walter, J. Dynamic link of DNA demethylation, DNA strand breaks and repair in mouse zygotes. *EMBO J.* **2010**, *29*, 1877–1888. [CrossRef] [PubMed]

37. Gehring, M.; Reik, W.; Henikoff, S. DNA demethylation by DNA repair. *Trends Genet.* **2009**, *25*, 82–90. [CrossRef]

38. Wu, H.; Zhang, Y. Reversing DNA Methylation: Mechanisms, Genomics, and Biological Functions. *Cell* **2014**, *156*, 45–68. [CrossRef]

39. Ladstätter, S.; Tachibana-Konwalski, K. A Surveillance Mechanism Ensures Repair of DNA Lesions during Zygotic Reprogramming. *Cell* **2016**, *167*, 1774–1787.e13. [CrossRef]

40. García-Rodríguez, A.; Gosálvez, J.; Agarwal, A.; Roy, R.; Johnston, S. DNA Damage and Repair in Human Reproductive Cells. *Int. J. Mol. Sci.* **2018**, *20*, 31. [CrossRef]

41. Smith, T.B.; Dun, M.D.; Smith, N.D.; Curry, B.J.; Connaughton, H.S.; Aitken, R.J. The presence of a truncated base excision repair pathway in human spermatozoa that is mediated by OGG1. *J. Cell Sci.* **2013**, *126*, 1488–1497. [CrossRef]

42. Borini, A.; Tarozzi, N.; Bizzaro, D.; Bonu, M.A.; Fava, L.; Flamigni, C.; Coticchio, G. Sperm DNA fragmentation: Paternal effect on early post-implantation embryo development in ART. *Hum. Reprod.* **2006**, *21*, 2876–2881. [CrossRef]

43. Rothkamm, K.; Krüger, I.; Thompson, L.H.; Löbrich, M. Pathways of DNA Double-Strand Break Repair during the Mammalian Cell Cycle. *MCB* **2003**, *23*, 5706–5715. [CrossRef] [PubMed]

44. Derijck, A.; van der Heijden, G.; Giele, M.; Philippens, M.; de Boer, P. DNA double-strand break repair in parental chromatin of mouse zygotes, the first cell cycle as an origin of de novo mutation. *Hum. Mol. Genet.* **2008**, *17*, 1922–1937. [CrossRef] [PubMed]

45. Osada, T.; Ogino, H.; Hino, T.; Ichinose, S.; Nakamura, K.; Omori, A.; Noce, T.; Masutani, M. PolyADP-Ribosylation Is Required for Pronuclear Fusion during Postfertilization in Mice. *PLoS ONE* **2010**, *5*, e12526. [CrossRef] [PubMed]

46. Izumi, T.; Mellon, I. Base Excision Repair and Nucleotide Excision Repair. In *Genome Stability*; Elsevier: Amsterdam, The Netherlands, 2016; pp. 275–302, ISBN 978-0-12-803309-8.

47. Tong, W.-M.; Cortes, U.; Wang, Z.-Q. Poly (ADP-ribose) polymerase: A guardian angel protecting the genome and suppressing tumorigenesis. *Biochim. Biophys. Acta BBA Rev. Cancer* **2001**, *1552*, 27–37. [CrossRef]

48. Gawecka, J.E.; Marh, J.; Ortega, M.; Yamauchi, Y.; Ward, M.A.; Ward, W.S. Mouse Zygotes Respond to Severe Sperm DNA Damage by Delaying Paternal DNA Replication and Embryonic Development. *PLoS ONE* **2013**, *8*, e56385. [CrossRef] [PubMed]

49. Adiga, S.K.; Toyoshima, M.; Shiraishi, K.; Shimura, T.; Takeda, J.; Taga, M.; Nagai, H.; Kumar, P.; Niwa, O. p21 provides stage specific DNA damage control to preimplantation embryos. *Oncogene* **2007**, *26*, 6141–6149. [CrossRef]

50. Bielanska, M.; Jin, S.; Bernier, M.; Tan, S.L.; Ao, A. Diploid-aneuploid mosaicism in human embryos cultured to the blastocyst stage. *Fertil. Steril.* **2005**, *84*, 336–342. [CrossRef]

51. Pfeiffer, P. Mechanisms of DNA double-strand break repair and their potential to induce chromosomal aberrations. *Mutagenesis* **2000**, *15*, 289–302. [CrossRef]

52. Petropoulos, S.; Panula, S.P.; Schell, J.P.; Lanner, F. Single-cell RNA sequencing: Revealing human pre-implantation development, pluripotency and germline development. *J. Intern. Med.* **2016**, *280*, 252–264. [CrossRef]

53. Lavagi, I.; Krebs, S.; Simmet, K.; Beck, A.; Zakhartchenko, V.; Wolf, E.; Blum, H. Single-cell RNA sequencing reveals developmental heterogeneity of blastomeres during major genome activation in bovine embryos. *Sci. Rep.* **2018**, *8*, 4071. [CrossRef]

54. Groff, A.F.; Resetkova, N.; DiDomenico, F.; Sakkas, D.; Penzias, A.; Rinn, J.L.; Eggan, K. RNA-seq as a tool for evaluating human embryo competence. *Genome Res.* **2019**, *29*, 1705–1718. [CrossRef] [PubMed]

55. Tichy, E.D.; Stambrook, P.J. DNA repair in murine embryonic stem cells and differentiated cells. *Exp. Cel Res.* **2008**, *314*, 1929–1936. [CrossRef] [PubMed]

56. Choi, E.-H.; Yoon, S.; Park, K.-S.; Kim, K.P. The Homologous Recombination Machinery Orchestrates Post-replication DNA Repair During Self-renewal of Mouse Embryonic Stem Cells. *Sci. Rep.* **2017**, *7*, 1–13. [CrossRef] [PubMed]

57. Ahuja, A.K.; Jodkowska, K.; Teloni, F.; Bizard, A.H.; Zellweger, R.; Herrador, R.; Ortega, S.; Hickson, I.D.; Altmeyer, M.; Mendez, J.; et al. A short G1 phase imposes constitutive replication stress and fork remodelling in mouse embryonic stem cells. *Nat. Commun.* **2016**, *7*, 10660. [CrossRef]

58. Van Sloun, P.P.H.; Jansen, J.G.; Weeda, G.; Mullenders, L.H.F.; van Zeeland, A.A.; Lohman, P.H.M.; Vrieling, H. The role of nucleotide excision repair in protecting embryonic stem cells from genotoxic effects of UV-induced DNA damage. *Nucleic Acids Res.* **1999**, *27*, 3276–3282. [CrossRef] [PubMed]

59. Kermi, C.; Aze, A.; Maiorano, D. Preserving Genome Integrity during the Early Embryonic DNA Replication Cycles. *Genes* **2019**, *10*, 398. [CrossRef]

60. Neganova, I.; Lako, M. G1 to S phase cell cycle transition in somatic and embryonic stem cells. *J. Anat.* **2008**, *213*, 30–44. [CrossRef]

61. Van der Laan, S.; Tsanov, N.; Crozet, C.; Maiorano, D. High Dub3 Expression in Mouse ESCs Couples the G1/S Checkpoint to Pluripotency. *Mol. Cell* **2013**, *52*, 366–379. [CrossRef]

62. Liu, L.; Michowski, W.; Kolodziejczyk, A.; Sicinski, P. The cell cycle in stem cell proliferation, pluripotency and differentiation. *Nat. Cell Biol.* **2019**, *21*, 1060–1067. [CrossRef]

63. Fu, X.; Cui, K.; Yi, Q.; Yu, L.; Xu, Y. DNA repair mechanisms in embryonic stem cells. *Cell. Mol. Life Sci.* **2017**, *74*, 487–493. [CrossRef]

Review

Exploring Mammalian Genome within Phase-Separated Nuclear Bodies: Experimental Methods and Implications for Gene Expression

Annick Lesne [1,2,*], Marie-Odile Baudement [1,3], Cosette Rebouissou [1] and Thierry Forné [1,*]

[1] IGMM, Univ. Montpellier, CNRS, F-34293 Montpellier, France; marie-odile.baudement@nmbu.no (M.-O.B.); cosette.rebouissou@igmm.cnrs.fr (C.R.)

[2] Sorbonne Université, CNRS, Laboratoire de Physique Théorique de la Matière Condensée, LPTMC, F-75252 Paris, France

[3] Centre for Integrative Genetics (CIGENE), Faculty of Biosciences, Norwegian University of Life Sciences, 1430 Ås, Norway

[*] Correspondence: annick.lesne@igmm.cnrs.fr (A.L.); forne@igmm.cnrs.fr (T.F.); Tel.: +33-434-359-682 (T.F.)

Received: 6 November 2019; Accepted: 13 December 2019; Published: 17 December 2019

Abstract: The importance of genome organization at the supranucleosomal scale in the control of gene expression is increasingly recognized today. In mammals, Topologically Associating Domains (TADs) and the active/inactive chromosomal compartments are two of the main nuclear structures that contribute to this organization level. However, recent works reviewed here indicate that, at specific loci, chromatin interactions with nuclear bodies could also be crucial to regulate genome functions, in particular transcription. They moreover suggest that these nuclear bodies are membrane-less organelles dynamically self-assembled and disassembled through mechanisms of phase separation. We have recently developed a novel genome-wide experimental method, High-salt Recovered Sequences sequencing (HRS-seq), which allows the identification of chromatin regions associated with large ribonucleoprotein (RNP) complexes and nuclear bodies. We argue that the physical nature of such RNP complexes and nuclear bodies appears to be central in their ability to promote efficient interactions between distant genomic regions. The development of novel experimental approaches, including our HRS-seq method, is opening new avenues to understand how self-assembly of phase-separated nuclear bodies possibly contributes to mammalian genome organization and gene expression.

Keywords: phase separation; nuclear bodies; self-assembly; genome organization; gene expression

1. Introduction

Several physical properties of nuclear organization are critical for regulating mammalian gene expression. In interphase, the genome is highly compacted to fit into the limited space of the cell nucleus while, at the same time, it remains fully accessible to multiple interactions involving *cis-* and *trans-*acting genomic elements and RNA/protein factors. Such a paradoxical achievement of a compact but dynamic genome is solved not only by packaging the genome into the chromatin nucleofilament, but also through a complex compartmentalization of the nucleus that contributes to the functional genome organization at the supranucleosomal scale (i.e., encompassing few tenths of kb to few Mb of DNA). The functional role of 3D genome organization has thus become an important component in the study of mammalian gene expression [1].

Another paradigm has been recently re-examined and developed: biomolecular condensates, grounded in the classical physical notion of phase separation [2]. While the use of this concept in a biological context dates back the old notion of coacervate, its relevance has been recently renewed by technological advances allowing in vivo observations and mechanistic investigations [3].

Phase separation describes the spontaneous formation of a two-phase system. From a physical point of view, it covers not only the demixing of oil and water, but also the spatial segregation that can arise in aqueous solutions, when the attraction between the solute molecules is energetically favored compared to the interaction between these molecules and the aqueous solvent. The balance between interaction energies and thermal motion or the ensuing diffusion, described by the free energy of the system, can lead in appropriate conditions to the spatial segregation of two phases of different concentrations [4]. This phenomenon is known as liquid–liquid phase separation (LLPS). Indeed, self-separated droplets display several features of a liquid phase: they are dense (as opposed to gases), display no rigid order (as opposed to crystals or liquid crystals), and their molecules remain mobile (as opposed to solids and gels), with permanent exchange between the two phases. These droplets display fluid-like behaviour, in particular the fusion of adjacent droplets into larger ones and a shape determined by surface tension. However, their composition, particularly under biological constraints, make them far more complicated than a mere liquid. Experimental strategies are thus developed to assess the presence and specificity of phase separation inside the cell [5]. Noticeably, "condensate" is the term used for molecular assemblies that form through phase separation while the more general term "hub" covers molecules that cluster together through yet unknown mechanisms.

Phase separation has been first recognized in the cytoplasm, as a mechanism of formation of stress granules and P-bodies [4]. It has been more recently invoked in the nucleus, for instance for the formation of membrane-less organelles also known as nuclear bodies. Much work is now devoted to identifying the hallmarks of in vivo phase separation and devising suitable protocols to study it [6]. In this review, we will first examine the proposal that nuclear compartments are phase-separated and could influence transcriptional regulation through their association with specific genomic sequences [7,8]. We will then present a novel experimental approach, HRS-seq, to test this working hypothesis.

2. Compartmentalization of Chromatin Interactions

In the past decade, the advent of sophisticated imaging techniques and molecular biology approaches based on proximity ligation assays (3C/Hi-C) has revealed that beyond the compaction achieved by packaging the DNA molecule at the nucleosomal level, chromatin is also organized within the three-dimensional (3D) space of the nucleus [9,10]. This 3D chromatin folding displays nested features, the most acknowledged being chromatin loops and topologically associating domains (TADs) where preferential *cis*-long-range contacts are observed [11]. A higher-order organization level also exists that partly covers the classic distinction between euchromatin and heterochromatin: the active (A) and inactive (B) chromosomal compartments [12]. While cohesin and CTCF proteins are required for TAD organization, these factors are dispensable for the maintenance of chromosomal compartments, which rely on different organization principle [13–15]. Furthermore, while TADs are essential for cell-specific genome organization and function [1], they appear to be quite stable between cell types, and even between organisms along evolution [16,17]. In striking contrast, chromatin loops and chromosomal compartments appear to vary during cell differentiation [18] and therefore they presumably play a central role for establishing specific gene expression profiles that determine cell identities. Several recent works started to decipher some crucial aspects of compartment regulation during mammalian spermatogenesis [19–22], in oocyte or early embryonic development [23–25], during cell differentiation [18,26] or reprograming [27] (for a recent review see [28]). However, to fully understand how 3D genome organization controls mammalian gene expression, it is critical to focus not only on long-range *cis*-interactions occurring at specific loci within TADs but also on *trans*-associations occurring between TADs within chromosomal compartments.

3. Nuclear Body Assembly by Phase Separation

Nuclear bodies are large membrane-less ribonucleoprotein (RNP) complexes known to be involved in several nuclear functions. For example, the synthesis of ribosomal RNAs (rRNAs) takes place in the nucleolus, the maturation of small nuclear RNAs (snRNAs) occurs in the Cajal bodies, and the histone messenger RNAs (mRNAs) are transcribed and matured in the histone-locus bodies (HLBs) (Table 1). One important aspect of functional nuclear compartmentalization is thus related to nuclear bodies. Some of them, like the HLBs, are known to gather loci that are dispersed in TADs located on distinct chromosomes, thus favouring coordinated gene transcription and efficient pre-mRNA maturation [29]. Similarly, the Cajal bodies have also been shown to contain inter-TAD interactions [30]. Transcription factories and active chromatin hubs are also large RNP complexes that have been proposed to coordinate gene expression by maintaining specific genes into a restricted 3D space of the nucleus [31]. Large RNP complexes, including some nuclear bodies, thus appear important for supranucleosomal genome organization in mammals. Indeed, their involvement in regulating transcription of specific genes suggests that they might be critical for the establishment and the maintenance of the active chromosomal compartment. However, the demonstration of such a role has so far been impeded by the lack of a genome-wide method that would allow unbiased profiling of genomic sequences associated with nuclear bodies. In our view, this is due to a continued lack of understanding of the physical nature of nuclear bodies in vivo.

Table 1. Classic nuclear bodies: main characteristics and components.

Compartment Name	Count/Nucleus	Diameter (μm)	Main Component	Main Associated Function	Ref.
Transcription factory	100	-	RNA Pol. II	mRNA transcription	[31]
Nucleolus	1–4	2–5	RNA Pol. I/Nucleolin	rRNA transcription	[32]
Cajal Body	10	0.5–1	Coilin, SMN [1]	Splicing	[33]
Gem	10	0.5–1	SMN1	SMN sequestration	[34]
Histone Locus Body	2–4	0.5–1	Coilin, NPAT [2]	Histone gene expression	[29]
Polycomb body	10–20	0.2–1.5	PRC1/PRC2 [3]	Histone PTMs [4]	[35]
PML body [5]	10–20	0.1–1	PML	Apoptosis, viral defense	[36]
Nuclear speckle	20–50	2–3	SC35 [6], RNA Pol. II	Splicing	[37,38]
Paraspeckle	10–20	0.5–1	NEAT1 [7] lncRNA	Transcription	[39,40]

[1] Survival of Motoneuron; [2] Nuclear protein, coactivator of histone transcription; [3] Polycomb repressive complexe; [4] PTM = Post-Translational Modifications; [5] Promyelocytic leukaemia nuclear body; [6] Serine/arginine-rich splicing factor; [7] Nuclear Paraspeckle Assembly Transcript 1.

It has been thought for a long time [41] that nuclear bodies are self-organized around nucleation sites, e.g., the Nucleolar Organizing Regions -NORs- for the nucleolus or the histone H3-H4 promoter region for Drosophila HLBs [42–44]. As a precedent, several cytoplasmic components, like the *Caenorhabditis elegans* P-granules [45] and centrosomes [46], have been discovered to behave in vivo like self-organized liquid-like droplets. More recently, based on in vitro experiments, other cytoplasmic structures, like the glycolytic bodies [47] or the RNA granules [48], have been proposed to form by phase separation processes. However, experimental evidence supporting self-organization or self-assembly remained very scarce for nuclear bodies (for reviews see [4,49]). A step forward has been the proposal, based on in vitro reconstitution experiments, that the phase separation of liquid-like RNP phases could control nucleolus size and assembly [50,51], as well as account for their sub-compartmentalized organization [52]. The demonstration that the Intrinsically Disordered Region (IDR) of Ddx4 protein (a critical component of the mammalian analogue to P-granules) can form phase-separated organelles, both in live cells and in vitro [53], led to the more precise hypothesis that phase separation of IDR-containing proteins could be a general mechanism for forming and regulating membrane-less organelles. These pioneering findings paved the way to a number of studies aimed at deciphering whether phase separation is involved in the organization of other nuclear compartments or bodies.

One can distinguish two phase separation processes: liquid–liquid phase separation (LLPS) and polymer–polymer phase separation (PPPS). While LLPS occurs through demixing of two liquid/liquid-like phases, PPPS involves bridging factors binding onto a polymer, e.g., the chromatin fiber, leading to a polymer collapse (i.e., a change in the polymer shape accompanied with an increase of its local density) [54]. Beyond the intrinsic nature of the interacting molecules responsible for phase separation (bridging factors for PPPS vs. weak multivalent binders for LLPS), the main differences between these two phase-separation processes lie in the role of the underlying polymer, if any. In PPPS, the polymer is required not only to nucleate phase separation but also to maintain it [55]. On the contrary, while an underlying polymer could help nucleation of LLPS, it is dispensable to maintain phase separation once a given saturating concentration of self-associating multivalent binder molecules has been reached [54].

Noticeably, phase separation was proposed to be involved in constitutive heterochromatin domain formation, based on the observation that a major component of the heterochromatin, the heterochromatin 1 α (HP1α) protein, can form liquid droplets both in vitro and in vivo [56,57]. HP1 self-oligomerization driven by phosphorylation is sufficient to induce HP1 phase separation in vitro [56]. However, since HP1α compartments can incorporate chromatin [56], the formation of heterochromatin domains in vivo, could actually be more complex [58] and rely not only on LLPS and weak multivalent chromatin binders [56,57], but also on PPPS, where a bridging factors, like the HP1 proteins themselves [59], could also induces a partial collapse of the chromatin [54,58].

4. Phase-Separation Models for Transcription Control

Following these discoveries, Phillip Sharp and colleagues proposed a phase-separation model for transcription control, in which a transcriptional multi-molecular assembly (i.e., a transcriptional condensate) would form by phase separation at a given locus following the formation of large RNP complexes induced by the binding of transcription factors at both enhancers and gene promoters [60]. This model was recently reinforced by studies showing that: (i) transcriptional coactivators, like BRD4 and the Mediator complex at active super-enhancers, together with the RNA polymerase II at promoters, can form transcriptional condensates in vitro [61,62], and (ii) domains driving gene activation in vivo are also required for phase separation in vitro [63]. Such transcriptional hubs, however, are relatively small compared to nuclear bodies. Therefore, it is not yet clear if their formation in vivo truly relies on phase separation and, if so, whether it is based on the demixing of two liquid-like phases similar to the LLPS observed for larger nuclear compartments like the nucleolus, or whether it reflects a hybrid situation also involving a polymer collapse process and PPPS as suspected in the case of heterochromatin domains. In all instances, we should remain careful before considering any transcriptional hub as a condensate formed by phase separation. Indeed, on the one hand RNA polymerase II was shown to form clusters or hubs at active genes through electrostatic interactions between its carboxy-terminal domain (CTD), a prominent IDR, and transcriptional coactivators, suggesting that compartmentalization may occur here through a LLPS process [7]. On the other hand, the transient unspecific binding of RNA polymerase II to the largely nucleosome free genome of the Herpes Simplex Virus type 1 (HSV1) leads to a DNA-mediated nuclear compartmentalization through a mechanism that is clearly distinct from LLPS [55]. Moreover, given the relatively small size of these transcriptional hubs, the physical properties that usually characterize the liquid state of the matter (like surface tension) may well make no real physical and biological sense [58,64]. That is precisely why the terms "hub" and "liquid-like phase separation" are often preferred to "condensate" and LLPS respectively [64]. The difference between a liquid and other states of the matter (like crystal, amorphous solid, liquid crystal or gel) lies in the mobility of the molecules, their ordered or disordered arrangement and the response to a stress (elastic vs. viscous). A whole range of intermediary behaviours are possible (e.g., the viscoelastic response of a gel). At the molecular scale, the liquid state is best characterized by the mobility of the molecules which is essentially depending on diffusion. Experimentally, FRAP (fluorescence recovery after photobleaching) experiments are used to quantitatively assess this mobility [50,57,63]. However, several caveats have been raised [5,6], the main one being that there are many physical models that can be fitted to the same fluorescence recovery curves [64]. Indeed, the rate of fluorescence recovery is not always due to freely diffusing molecules in solution, but could also depend on the local binding to others molecules. One critical point is thus to find experimental controls that could demonstrate that, independently of the models, the recovery rate is truly dominated by diffusion rather than binding.

It has been proposed that this could be achieved, for example, by showing a dependence of the recovery rate on the size of the bleach spot [65].

In parallel, another work indicated that various IDR-containing proteins form molecular hubs that could selectively associate in vivo with some chromatin regions by physically retaining targeted genomic loci while excluding non-targeted regions [66]. This chromatin filter model suggests that such molecular hubs could bring distal genomic loci together. However, these experiments use a novel CRISPR-Cas9-based technology (CasDrop) to artificially target chimeric IDR-containing proteins to chosen genomic sequences. It remains to be seen whether endogenous IDR-containing proteins act in a similar way on their natural targets. Additional work has shown the potential involvement of RNA-binding proteins [67].

In Figure 1, we provide an integrated model presenting the current working hypothesis, where we combine the concepts proposed in [60,63,66] for phase-separated transcriptional condensates involving long-range *cis*-interactions and extend these concepts to the probable involvement of nuclear bodies favoring inter-TADs *trans*-associations of co-regulated genes, like those observed for HLBs and Cajal bodies [30].

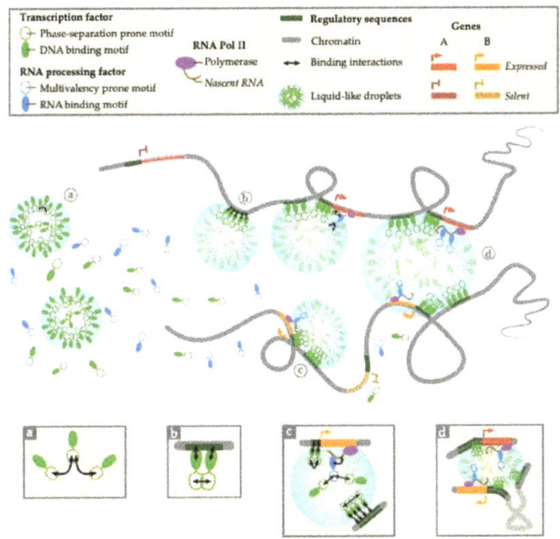

Figure 1. An integrated phase-separation model for self-assembly of transcriptional condensates controlling mammalian gene expression. (**a**) Transcription factors containing motifs prone to phase separation (e.g., IDR, Intrinsically Disordered Region) form liquid-like droplets (shaded in blue) by phase separation. (**b**) Their DNA binding motifs target specific genomic loci that are specifically incorporated into the droplets thus forming transcriptional condensates. Alternatively, phase separation could occur after binding of transcription factors (PPPS) on their target genomic sites in which case the corresponding DNA sequences act as nucleation sites. (**c**) Supplemented with the action of RNA processing factors containing motifs prone to multivalent interactions [67], they bring enhancers, promoters and/or nascent RNA transcripts in close vicinity, thus stabilizing long-range *cis*-interactions and promoting transcription. (**d**) In some instances, transcriptional condensates containing similar/compatible phase separation-prone motifs could finally merge into larger nuclear sub-organelles, leading to the formation of nuclear bodies like the Histone Locus Bodies (HLBs). The latter process brings together loci with similar transcriptional regulation but located on distinct (Topologically Associating Domains) TADs/chromosomes (orange/red lines and arrowheads), thus favouring the coordinated expression of the corresponding genes.

Our present working hypothesis, as synthesized in the integrated model (Figure 1), has two logical consequences: First, studying the physical principles and factors underlying the assembly of phase-separated nuclear bodies should differentiate at least two main classes of genes, those that are contacting phase-separated transcriptional condensates and those that are not, with as many sub-classes as types of condensates that interact with chromatin. Second, there should be at least two classes of membrane-less nuclear compartments, those that are depending (in vivo) on polymer–polymer phase separation (PPPS) and those that are depending on liquid–liquid or liquid-like phase separation (LLPS).

5. HRS-seq: A Novel Method to Explore Nuclear Bodies-Associated Sequences

Further exploration of the role of nuclear bodies in genome organization requires, as previously mentioned, an unbiased genome-wide sequencing of nuclear bodies-associated sequences. So far, these sequences have been difficult to analyse because no efficient and reliable method was available to purify nuclear bodies, presumably due to their membrane-less phase-separated nature.

It is known that performing high-salt treatments of transcriptionally active nuclei makes large RNP complexes, including nuclear bodies, insoluble [68]. More recently, we have shown that a 2M NaCl treatment traps the genomic DNA associated with these RNP complexes into the insoluble material which is easily purified on a filtration unit [69]. The trapped DNA fragments, that we named the "High-salt Recovered Sequences" (HRS), can then be separated from the rest of the genome by performing a simple restriction digestion and washing out the soluble material (Figure 2). The HRS thus remain on the filter unlike the rest of the genomic DNA. High-throughput sequencing of the HRS (HRS-seq) is then performed to obtain a global profiling of sequences associated with high-salt insoluble large RNP complexes, including nuclear bodies [70].

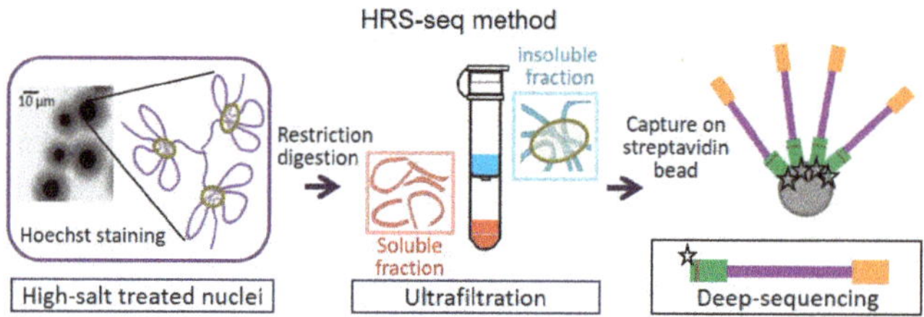

Figure 2. Principle of the HRS-seq method allowing the high-throughput identification of genomic sequences significantly associated with large RNP (ribonucleoprotein) complexes and nuclear bodies (adapted from [69]).

Most existing methods such as FAIRE-seq [71], ATAC-seq [72] or MNase-seq [73–75] aim at investigating accessibility of the chromatin nucleofilament at the nucleosomal scale. So far, only few approaches, like the HRS-seq, have been developed to investigate higher-order chromatin architecture at the supranucleosomal scale. Those include DamID mapping [76], 3C-derived methods like the Hi-C [12], MAR-seq [77] and TSA-seq [78]. Unlike all of these methods, the HRS-seq is avoiding delicate chemical crosslinking procedures or the use of specific antibodies that may restrict retention of some genomic sequences. Furthermore, it generally displays a better genomic resolution (few kb vs. few hundred kb) and is much straightforward and cheaper than existing approaches. However, in its present form, the HRS-seq method has several important limitations, the first of which is the fact that many large RNP complexes are extracted jointly in the insoluble material. A second limitation is that, contrary to 3C-derived approaches, it does not provide any indication on the physical proximity of the recovered sequences in vivo. Therefore, there is a clear need for improvements that would allow to identify sequences present simultaneously within specific subnuclear compartments. While assessing physical proximity will require to adapt a proximity-ligation assay to the HRS approach, the first limitation can already be addressed indirectly without modifying the existing HRS-seq protocol. Indeed, the inactivation of specific nuclear bodies by CRISPR/Cas9 technologies targeting critical components in relevant cellular models, combined with the present HRS-seq approach comparing wild-type and mutated cells, should soon allow extensive genomic profiling of sequences associated with specific nuclear bodies. This should lead to a much deeper understanding of how nuclear body-associated sequences and linked gene expression are dynamically affected during embryonic development and cellular differentiation, as well as in pathological situations where nuclear body formation is altered. For instance, in Spinal Muscular Atrophy (SMA), mutations of the *survival of motor neuron 1* (*SMN1*) gene affect Cajal bodies formation and lead to motor neuron death [79]. HRS-seq experiments on heathy or SMA-patient motor neurons should thus provide new insights on altered genomic organization and gene expression in the context of defective Cajal bodies.

The two logical consequences presented in the previous section could thus be tested in vivo using the HRS-seq method or quantitative PCR analyses of HRS assays (HRS-qPCR) in appropriate cellular models. Indeed, our recent work in mouse embryonic stem cells showed that HRS include sequences associated with nuclear bodies (like the Cajal bodies, the HLBs, the speckles and paraspeckles). Moreover, transcriptional hubs formed around super-enhancers are also retained in our assay [69]. In full agreement with the first consequence mentioned above, we found that two classes of genes can be defined according to the criterion of their association (or lack thereof) with large high-salt insoluble RNP complexes [69]. Our work showed that HRS-located genes are highly expressed and associated with the active chromosomal compartment and active super-enhancers in a cell-type specific manner, while genes that do not lie in HRS are moderately or weakly expressed.

Testing the second consequence will require experimental differentiation of PPPS from LLPS. As explained above, these two modalities of phase separation differ by the nature of the interacting molecules and the role played by the DNA/chromatin nucleofilament. Therefore, LLPS and liquid-like phase separation should be sensitive to compounds that disturb weak hydrophobic interactions, like moderate 1,6 hexanediol treatments [80], unlike PPPS that relies on stronger interactions. So far, sensitivity to 1,6 hexanediol has provided the best experimental evidence in favor of the involvement of liquid-like phase-separation processes in the assembly of transcriptional condensates in vivo [61,62], as well as for other classical nuclear bodies like the paraspeckles [81]. Therefore, combining 1,6 hexanediol treatments with HRS-seq could identify the genomic content of phase-separated condensates formed by LLPS driven by hydrophobic interactions. In contrast, molecular condensates formed by PPPS or those that would rely on a phase separation purely driven by electrostatic interactions (i.e., interactions between charged molecules, that are not disrupted by 1,6 hexanediol) are expected to be unaffected by 1,6 hexanediol treatment.

6. Discussion

The assembly of membrane-less compartments by phase separation appears to be a powerful mechanism for nuclear compartmentalization that could drive inter-TADs interactions between distant specific genomic loci. Such a compartmentalization could be essential to coordinate complex genomic functions, in particular transcription. At the molecular scale, thermal motion involved in phase-separation processes implies a continuous exchange of molecules between the dense and the dilute phases. Phase separation depends on the local concentration within the nucleus (or a region of the nucleus) of critical components, like IDR-containing proteins, and can thus be controlled by regulating their availability. This could be achieved by simple post-translational modifications that affect the protein's ability to establish multivalent interactions, like phosphorylation [82]. Supporting this, PRKACB (catalytic subunit of PKA cAMP-dependent protein kinase) and HIP (Homeodomain-interacting protein) kinases are required for the in vivo assembly of the Cajal and PML bodies, respectively [83]. However, little is known about nuclear body homeostasis, which certainly constitutes a promising topic for future investigations.

Liquid–liquid phase separation is not a feature involving an isolated molecular species but is depending on the properties of both this species and the solvent. In the case of nuclear bodies, the solvent corresponds to the complex nuclear environment in which the molecular species of interest is considered. A modification of the contents of this environment would thus affect phase separation. Obviously, a structural or chemical modification of the phase separation-prone molecular species would also affect spatial structuration. Various means of tuning the physical process of phase separation are thus possible within a living cell. While in vitro experiments usually monitor physical parameters controlling phase separation (like temperature or pH), in vivo a specific adaptation of the relative strength of the molecular interactions, through some post-translational modification of the phase separation-prone protein, would offer a more precise control of the process.

The current thermodynamic description of phase separation processes is only valid on a large scale (i.e., involving large enough number of molecules). The direct effects of the intrinsic stochasticity prevailing at molecular scales are random binding/unbinding of interacting molecules, diffusion and ensuing concentration fluctuations. They are included only in an average way in the large-scale thermodynamic description. In case of small systems with a finite number of molecules, (e.g., a region of the nucleus), discrepancies may arise, among which a modification of the stability regions in the parameter space, loss of correlations in cooperative assembly, or various diffusion-limited behaviours. Thus, the robustness with respect to molecular noise of a thermodynamically predicted phase separation needs to be investigated. In the spirit of studies quantifying the stochasticity of transcription [84], the analysis of imaging data or measurements obtained from a large number of single cells observed in the same conditions would assess the variability of the phase separation phenomenon. On the theoretical side, the thermodynamic approach could be supplemented with stochastic dynamical equations including a noise term [85] and the simulation of their solutions [60,86].

Finally, to date, investigations have relied on the description of phase separation in the framework of thermodynamic equilibrium. Nevertheless, active processes are possibly at work in vivo. An example is the observation of droplet fission [87] that is not accounted for in the current thermodynamic models of phase separation. Investigating active features of intracellular dynamic organization thus opens a fascinating research field not only for biologists but also for theoretical physicists. Phase separation is actually a special instance of the more general concept of self-organization, in which a long-range spatial structuring emerges from short-range interactions and breaks the symmetry of the homogenous state. The mechanisms underlying self-organization range from self-assembly of equilibrium complexes to out-of-equilibrium formation of dissipative structures [88,89]. It is thus plausible that a variety of different mechanisms could be involved inside the cell.

7. Conclusions

The physical notion of phase separation opens novel research avenues in the field of transcriptional gene regulation by suggesting a possible interplay between assembly of nuclear bodies and recruitment of specific genomic sequences. However, it remains to be determined to what extent such interplay is dependent on phase separation, or on more complex active and/or specific processes. Here, HRS-seq, combined with other approaches, can be instrumental for dissecting the relationship between 3D chromatin organization and nuclear bodies, and its implication for both *cis-* and *trans-* co-regulation of gene expression. Understanding the relevance of phase separation in a biological context will require theoretical studies devising microscopic descriptions accounting for the intrinsic fluctuations present at the intracellular scale, as well as experimental studies investigating the possible involvement of active mechanisms.

Author Contributions: All authors contributed to the conception and writing of this manuscript.

Funding: This research was funded by the A.F.M.-Téléthon grant number 21024, *ANR* CHRODYT grant number ANR-16-CE15–0018–04, the Cancéropole GSO grant number 2018-E08 and the C.N.R.S. (Centre National de la Recherche Scientifique).

Acknowledgments: We thank G. Cathala for stimulating discussions, R. Hipskind for his critical reading of the manuscript, and A. Taudière (IdEst) for the preparation of Figure 1.

Conflicts of Interest: The authors declare no conflict of interest.

Abbreviations

3D	three-dimensional
HLB	histone locus body
HRS	High-salt Recovered Sequences
IDR	intrinsically disordered region
LLPS	liquid–liquid phase separation
PPPS	polymer–polymer phase separation
RNP	ribonucleoprotein
TAD	topologically associating domain
TF	transcription factor

References

1. Pombo, A.; Dillon, N. Three-dimensional genome architecture: Players and mechanisms. *Nat. Rev. Mol. Cell Biol.* **2015**, *16*, 245–257. [CrossRef] [PubMed]
2. Banani, S.F.; Lee, H.O.; Hyman, A.A.; Rosen, M.K. Biomolecular condensates: Organizers of cellular biochemistry. *Nat. Rev. Mol. Cell Biol.* **2017**, *18*, 285–298. [CrossRef] [PubMed]
3. Alberti, S. Phase separation in biology. *Curr. Biol.* **2017**, *27*, R1097–R1102. [CrossRef] [PubMed]
4. Hyman, A.A.; Weber, C.A.; Julicher, F. Liquid-liquid phase separation in biology. *Annu. Rev. Cell Dev. Biol.* **2014**, *30*, 39–58. [CrossRef]
5. Alberti, S.; Gladfelter, A.; Mittag, T. Considerations and Challenges in Studying Liquid-Liquid Phase Separation and Biomolecular Condensates. *Cell* **2019**, *176*, 419–434. [CrossRef]
6. Mir, M.; Bickmore, W.; Furlong, E.E.M.; Narlikar, G. Chromatin topology, condensates and gene regulation: Shifting paradigms or just a phase? *Development* **2019**, *146*, dev182766. [CrossRef]
7. Boehning, M.; Dugast-Darzacq, C.; Rankovic, M.; Hansen, A.S.; Yu, T.; Marie-Nelly, H.; McSwiggen, D.T.; Kokic, G.; Dailey, G.M.; Cramer, P.; et al. RNA polymerase II clustering through carboxy-terminal domain phase separation. *Nat. Struct. Mol. Biol.* **2018**, *25*, 833–840. [CrossRef]
8. Chong, S.; Dugast-Darzacq, C.; Liu, Z.; Dong, P.; Dailey, G.M.; Cattoglio, C.; Heckert, A.; Banala, S.; Lavis, L.; Darzacq, X.; et al. Imaging dynamic and selective low-complexity domain interactions that control gene transcription. *Science* **2018**, *361*, aar2555. [CrossRef]
9. Court, F.; Miro, J.; Braem, C.; Lelay-Taha, M.N.; Brisebarre, A.; Atger, F.; Gostan, T.; Weber, M.; Cathala, G.; Forne, T. Modulated contact frequencies at gene-rich loci support a statistical helix model for mammalian chromatin organization. *Genome Biol.* **2011**, *12*, R42. [CrossRef]
10. Ea, V.; Baudement, M.O.; Lesne, A.; Forné, T. Contribution of Topological Domains and Loop Formation to 3D Chromatin Organization. *Genes* **2015**, *6*, 734–750. [CrossRef]
11. Nora, E.P.; Lajoie, B.R.; Schulz, E.G.; Giorgetti, L.; Okamoto, I.; Servant, N.; Piolot, T.; van Berkum, N.L.; Meisig, J.; Sedat, J.; et al. Spatial partitioning of the regulatory landscape of the X-inactivation centre. *Nature* **2012**, *485*, 381–385. [CrossRef] [PubMed]
12. Lieberman-Aiden, E.; van Berkum, N.L.; Williams, L.; Imakaev, M.; Ragoczy, T.; Telling, A.; Amit, I.; Lajoie, B.R.; Sabo, P.J.; Dorschner, M.O.; et al. Comprehensive mapping of long-range interactions reveals folding principles of the human genome. *Science* **2009**, *326*, 289–293. [CrossRef] [PubMed]
13. Nora, E.P.; Goloborodko, A.; Valton, A.L.; Gibcus, J.H.; Uebersohn, A.; Abdennur, N.; Dekker, J.; Mirny, L.A.; Bruneau, B.G. Targeted Degradation of CTCF Decouples Local Insulation of Chromosome Domains from Genomic Compartmentalization. *Cell* **2017**, *169*, 930–944. [CrossRef] [PubMed]
14. Rao, S.S.P.; Huang, S.C.; Glenn St Hilaire, B.; Engreitz, J.M.; Perez, E.M.; Kieffer-Kwon, K.R.; Sanborn, A.L.; Johnstone, S.E.; Bascom, G.D.; Bochkov, I.D.; et al. Cohesin Loss Eliminates All Loop Domains. *Cell* **2017**, *171*, 305–320. [CrossRef] [PubMed]
15. Schwarzer, W.; Abdennur, N.; Goloborodko, A.; Pekowska, A.; Fudenberg, G.; Loe-Mie, Y.; Fonseca, N.A.; Huber, W.; Haering, C.H.; Mirny, L.; et al. Two independent modes of chromatin organization revealed by cohesin removal. *Nature* **2017**, *551*, 51–56. [CrossRef] [PubMed]
16. Dixon, J.R.; Jung, I.; Selvaraj, S.; Shen, Y.; Antosiewicz-Bourget, J.E.; Lee, A.Y.; Ye, Z.; Kim, A.; Rajagopal, N.; Xie, W.; et al. Chromatin architecture reorganization during stem cell differentiation. *Nature* **2015**, *518*, 331–336. [CrossRef]

17. Dixon, J.R.; Selvaraj, S.; Yue, F.; Kim, A.; Li, Y.; Shen, Y.; Hu, M.; Liu, J.S.; Ren, B. Topological domains in mammalian genomes identified by analysis of chromatin interactions. *Nature* **2012**, *485*, 376–380. [CrossRef]

18. Bonev, B.; Mendelson Cohen, N.; Szabo, Q.; Fritsch, L.; Papadopoulos, G.L.; Lubling, Y.; Xu, X.; Lv, X.; Hugnot, J.P.; Tanay, A.; et al. Multiscale 3D Genome Rewiring during Mouse Neural Development. *Cell* **2017**, *171*, 557–572. [CrossRef]

19. Alavattam, K.G.; Maezawa, S.; Sakashita, A.; Khoury, H.; Barski, A.; Kaplan, N.; Namekawa, S.H. Attenuated chromatin compartmentalization in meiosis and its maturation in sperm development. *Nat. Struct. Mol. Biol.* **2019**, *26*, 175–184. [CrossRef]

20. Gassler, J.; Brandao, H.B.; Imakaev, M.; Flyamer, I.M.; Ladstatter, S.; Bickmore, W.A.; Peters, J.M.; Mirny, L.A.; Tachibana, K. A mechanism of cohesin-dependent loop extrusion organizes zygotic genome architecture. *EMBO J.* **2017**, *36*, 3600–3618. [CrossRef]

21. Patel, L.; Kang, R.; Rosenberg, S.C.; Qiu, Y.; Raviram, R.; Chee, S.; Hu, R.; Ren, B.; Cole, F.; Corbett, K.D. Dynamic reorganization of the genome shapes the recombination landscape in meiotic prophase. *Nat. Struct. Mol. Biol.* **2019**, *26*, 164–174. [CrossRef] [PubMed]

22. Wang, Y.; Wang, H.; Zhang, Y.; Du, Z.; Si, W.; Fan, S.; Qin, D.; Wang, M.; Duan, Y.; Li, L.; et al. Reprogramming of Meiotic Chromatin Architecture during Spermatogenesis. *Mol. Cell* **2019**, *73*, 547–561. [CrossRef] [PubMed]

23. Du, Z.; Zheng, H.; Huang, B.; Ma, R.; Wu, J.; Zhang, X.; He, J.; Xiang, Y.; Wang, Q.; Li, Y.; et al. Allelic reprogramming of 3D chromatin architecture during early mammalian development. *Nature* **2017**, *547*, 232–235. [CrossRef] [PubMed]

24. Flyamer, I.M.; Gassler, J.; Imakaev, M.; Brandao, H.B.; Ulianov, S.V.; Abdennur, N.; Razin, S.V.; Mirny, L.A.; Tachibana-Konwalski, K. Single-nucleus Hi-C reveals unique chromatin reorganization at oocyte-to-zygote transition. *Nature* **2017**, *544*, 110–114. [CrossRef]

25. Ke, Y.; Xu, Y.; Chen, X.; Feng, S.; Liu, Z.; Sun, Y.; Yao, X.; Li, F.; Zhu, W.; Gao, L.; et al. 3D Chromatin Structures of Mature Gametes and Structural Reprogramming during Mammalian Embryogenesis. *Cell* **2017**, *170*, 367–381. [CrossRef]

26. Miura, H.; Takahashi, S.; Poonperm, R.; Tanigawa, A.; Takebayashi, S.I.; Hiratani, I. Single-cell DNA replication profiling identifies spatiotemporal developmental dynamics of chromosome organization. *Nat. Genet.* **2019**, *51*, 1356–1368. [CrossRef]

27. Stadhouders, R.; Vidal, E.; Serra, F.; Di Stefano, B.; Le Dily, F.; Quilez, J.; Gomez, A.; Collombet, S.; Berenguer, C.; Cuartero, Y.; et al. Transcription factors orchestrate dynamic interplay between genome topology and gene regulation during cell reprogramming. *Nat. Genet.* **2018**, *50*, 238–249. [CrossRef]

28. Zheng, H.; Xie, W. The role of 3D genome organization in development and cell differentiation. *Nat. Rev. Mol. Cell Biol.* **2019**, *20*, 535–550. [CrossRef]

29. Frey, M.R.; Matera, A.G. Coiled bodies contain U7 small nuclear RNA and associate with specific DNA sequences in interphase human cells. *Proc. Natl. Acad. Sci. USA* **1995**, *92*, 5915–5919. [CrossRef]

30. Wang, Q.; Sawyer, I.A.; Sung, M.H.; Sturgill, D.; Shevtsov, S.P.; Pegoraro, G.; Hakim, O.; Baek, S.; Hager, G.L.; Dundr, M. Cajal bodies are linked to genome conformation. *Nat. Commun.* **2016**, *7*, 10966. [CrossRef]

31. Jackson, D.A.; Iborra, F.J.; Manders, E.M.; Cook, P.R. Numbers and organization of RNA polymerases, nascent transcripts, and transcription units in HeLa nuclei. *Mol. Biol. Cell* **1998**, *9*, 1523–1536. [CrossRef] [PubMed]

32. Boisvert, F.M.; van Koningsbruggen, S.; Navascues, J.; Lamond, A.I. The multifunctional nucleolus. *Nat. Rev. Mol. Cell Biol.* **2007**, *8*, 574–585. [CrossRef] [PubMed]

33. Gall, J.G. The centennial of the Cajal body. *Nat. Rev. Mol. Cell Biol.* **2003**, *4*, 975–980. [CrossRef] [PubMed]

34. Liu, Q.; Dreyfuss, G. A novel nuclear structure containing the survival of motor neurons protein. *EMBO J.* **1996**, *15*, 3555–3565. [CrossRef]

35. Saurin, A.J.; Shiels, C.; Williamson, J.; Satijn, D.P.; Otte, A.P.; Sheer, D.; Freemont, P.S. The human polycomb group complex associates with pericentromeric heterochromatin to form a novel nuclear domain. *J. Cell Biol.* **1998**, *142*, 887–898. [CrossRef]

36. Lallemand-Breitenbach, V.; de The, H. PML nuclear bodies. *Cold Spring Harb. Perspect. Biol.* **2010**, *2*, a000661. [CrossRef]

37. Hutchinson, J.N.; Ensminger, A.W.; Clemson, C.M.; Lynch, C.R.; Lawrence, J.B.; Chess, A. A screen for nuclear transcripts identifies two linked noncoding RNAs associated with SC35 splicing domains. *BMC Genom.* **2007**, *8*, 39. [CrossRef]

38. Saitoh, N.; Spahr, C.S.; Patterson, S.D.; Bubulya, P.; Neuwald, A.F.; Spector, D.L. Proteomic analysis of interchromatin granule clusters. *Mol. Biol. Cell* **2004**, *15*, 3876–3890. [CrossRef]

39. Clemson, C.M.; Hutchinson, J.N.; Sara, S.A.; Ensminger, A.W.; Fox, A.H.; Chess, A.; Lawrence, J.B. An architectural role for a nuclear noncoding RNA: NEAT1 RNA is essential for the structure of paraspeckles. *Mol. Cell* **2009**, *33*, 717–726. [CrossRef]

40. Naganuma, T.; Nakagawa, S.; Tanigawa, A.; Sasaki, Y.F.; Goshima, N.; Hirose, T. Alternative 3′-end processing of long noncoding RNA initiates construction of nuclear paraspeckles. *EMBO J.* **2012**, *31*, 4020–4034. [CrossRef]

41. Misteli, T. The concept of self-organization in cellular architecture. *J. Cell Biol.* **2001**, *155*, 181–185. [CrossRef] [PubMed]

42. Grob, A.; Colleran, C.; McStay, B. Construction of synthetic nucleoli in human cells reveals how a major functional nuclear domain is formed and propagated through cell division. *Genes Dev.* **2014**, *28*, 220–230. [CrossRef] [PubMed]

43. Kaiser, T.E.; Intine, R.V.; Dundr, M. De novo formation of a subnuclear body. *Science* **2008**, *322*, 1713–1717. [CrossRef] [PubMed]

44. Salzler, H.R.; Tatomer, D.C.; Malek, P.Y.; McDaniel, S.L.; Orlando, A.N.; Marzluff, W.F.; Duronio, R.J. A sequence in the Drosophila H3-H4 Promoter triggers histone locus body assembly and biosynthesis of replication-coupled histone mRNAs. *Dev. Cell* **2013**, *24*, 623–634. [CrossRef] [PubMed]

45. Brangwynne, C.P.; Eckmann, C.R.; Courson, D.S.; Rybarska, A.; Hoege, C.; Gharakhani, J.; Julicher, F.; Hyman, A.A. Germline P granules are liquid droplets that localize by controlled dissolution/condensation. *Science* **2009**, *324*, 1729–1732. [CrossRef] [PubMed]

46. Zwicker, D.; Decker, M.; Jaensch, S.; Hyman, A.A.; Julicher, F. Centrosomes are autocatalytic droplets of pericentriolar material organized by centrioles. *Proc. Natl. Acad. Sci. USA* **2014**, *111*, E2636–E2645. [CrossRef]

47. Jin, M.; Fuller, G.G.; Han, T.; Yao, Y.; Alessi, A.F.; Freeberg, M.A.; Roach, N.P.; Moresco, J.J.; Karnovsky, A.; Baba, M.; et al. Glycolytic Enzymes Coalesce in G Bodies under Hypoxic Stress. *Cell Rep.* **2017**, *20*, 895–908. [CrossRef]

48. Kato, M.; Han, T.W.; Xie, S.; Shi, K.; Du, X.; Wu, L.C.; Mirzaei, H.; Goldsmith, E.J.; Longgood, J.; Pei, J.; et al. Cell-free formation of RNA granules: Low complexity sequence domains form dynamic fibers within hydrogels. *Cell* **2012**, *149*, 753–767. [CrossRef]

49. Zhu, L.; Brangwynne, C.P. Nuclear bodies: The emerging biophysics of nucleoplasmic phases. *Curr. Opin. Cell. Biol.* **2015**, *34*, 23–30. [CrossRef]

50. Berry, J.; Weber, S.C.; Vaidya, N.; Haataja, M.; Brangwynne, C.P. RNA transcription modulates phase transition-driven nuclear body assembly. *Proc. Natl. Acad. Sci. USA* **2015**, *112*, E5237–E5245. [CrossRef]

51. Weber, S.C.; Brangwynne, C.P. Inverse size scaling of the nucleolus by a concentration-dependent phase transition. *Curr. Biol.* **2015**, *25*, 641–646. [CrossRef] [PubMed]

52. Feric, M.; Vaidya, N.; Harmon, T.S.; Mitrea, D.M.; Zhu, L.; Richardson, T.M.; Kriwacki, R.W.; Pappu, R.V.; Brangwynne, C.P. Coexisting Liquid Phases Underlie Nucleolar Subcompartments. *Cell* **2016**, *165*, 1686–1697. [CrossRef] [PubMed]

53. Nott, T.J.; Petsalaki, E.; Farber, P.; Jervis, D.; Fussner, E.; Plochowietz, A.; Craggs, T.D.; Bazett-Jones, D.P.; Pawson, T.; Forman-Kay, J.D.; et al. Coexisting Liquid Phases Underlie Nucleolar Subcompartments. *Cell* **2015**, *57*, 936–947.

54. Erdel, F.; Rippe, K. Formation of Chromatin Subcompartments by Phase Separation. *Biophys. J.* **2018**, *114*, 2262–2270. [CrossRef] [PubMed]

55. McSwiggen, D.T.; Hansen, A.S.; Teves, S.S.; Marie-Nelly, H.; Hao, Y.; Heckert, A.B.; Umemoto, K.K.; Dugast-Darzacq, C.; Tjian, R.; Darzacq, X. Evidence for DNA-mediated nuclear compartmentalization distinct from phase separation. *Elife* **2019**, *8*, 47098. [CrossRef] [PubMed]

56. Larson, A.G.; Elnatan, D.; Keenen, M.M.; Trnka, M.J.; Johnston, J.B.; Burlingame, A.L.; Agard, D.A.; Redding, S.; Narlikar, G.J. Liquid droplet formation by HP1alpha suggests a role for phase separation in heterochromatin. *Nature* **2017**, *547*, 236–240. [CrossRef]

57. Strom, A.R.; Emelyanov, A.V.; Mir, M.; Fyodorov, D.V.; Darzacq, X.; Karpen, G.H. Phase separation drives heterochromatin domain formation. *Nature* **2017**, *547*, 241–245. [CrossRef]

58. Peng, A.; Weber, S.C. Evidence for and against Liquid-Liquid Phase Separation in the Nucleus. *Noncoding RNA.* **2019**, *5*, ncrna5040050.

59. Machida, S.; Takizawa, Y.; Ishimaru, M.; Sugita, Y.; Sekine, S.; Nakayama, J.I.; Wolf, M.; Kurumizaka, H. Structural Basis of Heterochromatin Formation by Human HP1. *Mol. Cell* **2018**, *69*, 385–397.e388. [CrossRef]

60. Hnisz, D.; Shrinivas, K.; Young, R.A.; Chakraborty, A.K.; Sharp, P.A. A Phase Separation Model for Transcriptional Control. *Cell* **2017**, *169*, 13–23. [CrossRef]

61. Cho, W.K.; Spille, J.H.; Hecht, M.; Lee, C.; Li, C.; Grube, V.; Cisse, I.I. Mediator and RNA polymerase II clusters associate in transcription-dependent condensates. *Science* **2018**, *361*, 412–415. [CrossRef] [PubMed]

62. Sabari, B.R.; Dall'Agnese, A.; Boija, A.; Klein, I.A.; Coffey, E.L.; Shrinivas, K.; Abraham, B.J.; Hannett, N.M.; Zamudio, A.V.; Manteiga, J.C.; et al. Coactivator condensation at super-enhancers links phase separation and gene control. *Science* **2018**, *361*, aar3958. [CrossRef] [PubMed]

63. Boija, A.; Klein, I.A.; Sabari, B.R.; Dall'Agnese, A.; Coffey, E.L.; Zamudio, A.V.; Li, C.H.; Shrinivas, K.; Manteiga, J.C.; Hannett, N.M.; et al. Transcription Factors Activate Genes through the Phase-Separation Capacity of Their Activation Domains. *Cell* **2018**, *175*, 1842–1855. [CrossRef] [PubMed]

64. McSwiggen, D.T.; Mir, M.; Darzacq, X.; Tjian, R. Evaluating phase separation in live cells: Diagnosis, caveats, and functional consequences. *Genes Dev.* **2019**, *8*, 119. [CrossRef] [PubMed]

65. Taylor, N.O.; Wei, M.T.; Stone, H.A.; Brangwynne, C.P. Quantifying Dynamics in Phase-Separated Condensates Using Fluorescence Recovery after Photobleaching. *Biophys. J.* **2019**, *117*, 1285–1300. [CrossRef] [PubMed]

66. Shin, Y.; Chang, Y.C.; Lee, D.S.W.; Berry, J.; Sanders, D.W.; Ronceray, P.; Wingreen, N.S.; Haataja, M.; Brangwynne, C.P. Liquid Nuclear Condensates Mechanically Sense and Restructure the Genome. *Cell* **2018**, *175*, 1481–1491. [CrossRef]

67. Lin, Y.; Protter, D.S.; Rosen, M.K.; Parker, R. Formation and Maturation of Phase-Separated Liquid Droplets by RNA-Binding Proteins. *Mol. Cell* **2015**, *60*, 208–219. [CrossRef]

68. Engelke, R.; Riede, J.; Hegermann, J.; Wuerch, A.; Eimer, S.; Dengjel, J.; Mittler, G. The quantitative nuclear matrix proteome as a biochemical snapshot of nuclear organization. *J. Proteome Res.* **2014**, *13*, 3940–3956. [CrossRef]

69. Baudement, M.O.; Cournac, A.; Court, F.; Seveno, M.; Parrinello, H.; Reynes, C.; Sabatier, R.; Bouschet, T.; Yi, Z.; Sallis, S.; et al. High-salt-recovered sequences are associated with the active chromosomal compartment and with large ribonucleoprotein complexes including nuclear bodies. *Genome Res.* **2018**, *28*, 1733–1746. [CrossRef]

70. Braem, C.; Recolin, B.; Rancourt, R.C.; Angiolini, C.; Barthes, P.; Branchu, P.; Court, F.; Cathala, G.; Ferguson-Smith, A.C.; Forné, T. Genomic matrix attachment region and chromosome conformation capture quantitative real time PCR assays identify novel putative regulatory elements at the imprinted Dlk1/Gtl2 locus. *J. Biol. Chem.* **2008**, *283*, 18612–18620. [CrossRef]

71. Giresi, P.G.; Kim, J.; McDaniell, R.M.; Iyer, V.R.; Lieb, J.D. FAIRE (Formaldehyde-Assisted Isolation of Regulatory Elements) isolates active regulatory elements from human chromatin. *Genome Res.* **2007**, *17*, 877–885. [CrossRef] [PubMed]

72. Buenrostro, J.D.; Giresi, P.G.; Zaba, L.C.; Chang, H.Y.; Greenleaf, W.J. Transposition of native chromatin for fast and sensitive epigenomic profiling of open chromatin, DNA-binding proteins and nucleosome position. *Nat. Methods* **2013**, *10*, 1213–1218. [CrossRef] [PubMed]

73. Gaffney, D.J.; McVicker, G.; Pai, A.A.; Fondufe-Mittendorf, Y.N.; Lewellen, N.; Michelini, K.; Widom, J.; Gilad, Y.; Pritchard, J.K. Controls of nucleosome positioning in the human genome. *PLoS Genet.* **2012**, *8*, e1003036. [CrossRef] [PubMed]

74. Henikoff, J.G.; Belsky, J.A.; Krassovsky, K.; MacAlpine, D.M.; Henikoff, S. Epigenome characterization at single base-pair resolution. *Proc. Natl. Acad. Sci. USA* **2011**, *108*, 18318–18323. [CrossRef]

75. Valouev, A.; Johnson, S.M.; Boyd, S.D.; Smith, C.L.; Fire, A.Z.; Sidow, A. Determinants of nucleosome organization in primary human cells. *Nature* **2011**, *474*, 516–520. [CrossRef]

76. Vogel, M.J.; Peric-Hupkes, D.; van Steensel, B. Detection of in vivo protein-DNA interactions using DamID in mammalian cells. *Nat. Protoc.* **2007**, *2*, 1467–1478. [CrossRef]

77. Dobson, J.R.; Hong, D.; Barutcu, A.R.; Wu, H.; Imbalzano, A.N.; Lian, J.B.; Stein, J.L.; van Wijnen, A.J.; Nickerson, J.A.; Stein, G.S. Identifying Nuclear Matrix-Attached DNA Across the Genome. *J. Cell. Physiol.* **2017**, *232*, 1295–1305. [CrossRef]

78. Chen, Y.; Zhang, Y.; Wang, Y.; Zhang, L.; Brinkman, E.K.; Adam, S.A.; Goldman, R.; van Steensel, B.; Ma, J.; Belmont, A.S. Mapping 3D genome organization relative to nuclear compartments using TSA-Seq as a cytological ruler. *J. Cell Biol.* **2018**, *217*, 4025–4048. [CrossRef]

79. Zhang, R.; So, B.R.; Li, P.; Yong, J.; Glisovic, T.; Wan, L.; Dreyfuss, G. Structure of a key intermediate of the SMN complex reveals Gemin2's crucial function in snRNP assembly. *Cell* **2011**, *146*, 384–395. [CrossRef]

80. Kroschwald, S.; Maharana, S.; Alberti, S. Hexanediol: A chemical probe to investigate the material properties of membrane-less compartments. *Matters* **2017**. [CrossRef]

81. Yamazaki, T.; Souquere, S.; Chujo, T.; Kobelke, S.; Chong, Y.S.; Fox, A.H.; Bond, C.S.; Nakagawa, S.; Pierron, G.; Hirose, T. Functional Domains of NEAT1 Architectural lncRNA Induce Paraspeckle Assembly through Phase Separation. *Mol. Cell* **2018**, *70*, 1038–1053. [CrossRef] [PubMed]

82. Hebert, M.D.; Poole, A.R. Towards an understanding of regulating Cajal body activity by protein modification. *RNA Biol.* **2017**, *14*, 761–778. [CrossRef] [PubMed]

83. Berchtold, D.; Battich, N.; Pelkmans, L. A Systems-Level Study Reveals Regulators of Membrane-less Organelles in Human Cells. *Mol. Cell* **2018**, *72*, 1035–1049. [CrossRef] [PubMed]

84. Elowitz, M.B.; Levine, A.J.; Siggia, E.D.; Swain, P.S. Stochastic gene expression in a single cell. *Science* **2002**, *297*, 1183–1186. [CrossRef]

85. Berry, J.; Brangwynne, C.P.; Haataja, M. Physical principles of intracellular organization via active and passive phase transitions. *Rep. Prog. Phys.* **2018**, *81*, 046601. [CrossRef]

86. Lemarchand, A.; Lesne, A.; Mareschal, M. Langevin approach to a chemical wave front: Selection of the propagation velocity in the presence of internal noise. *Phys. Rev. E* **1995**, *51*, 4457–4465. [CrossRef]

87. Dellaire, G.; Ching, R.W.; Dehghani, H.; Ren, Y.; Bazett-Jones, D.P. The number of PML nuclear bodies increases in early S phase by a fission mechanism. *J. Cell Sci.* **2006**, *119*, 1026–1033. [CrossRef]

88. Camazine, S.; Deneubourg, J.L.; Franks, N.R.; Sneyd, J.; Bonabeau, E.; Theraulaz, G. *Self-Organization in Biological Systems*; Princeton University Press: Princeton, NJ, USA, 2003.

89. Lehn, J.M. Toward complex matter: Supramolecular chemistry and self-organization. *Proc. Natl. Acad. Sci. USA* **2002**, *99*, 4763–4768. [CrossRef]

MDPI

St. Alban-Anlage 66

4052 Basel

Switzerland

Tel. +41 61 683 77 34

Fax +41 61 302 89 18

www.mdpi.com

Genes Editorial Office

E-mail: genes@mdpi.com

www.mdpi.com/journal/genes